Introduction to Construction Drawing

Arthur Thompson
MBIAT, Associate Member CIOB, Cert. Ed.
Formerly Senior Lecturer, Vauxhall College, London

Edward Arnold
A division of Hodder & Stoughton
LONDON MELBOURNE AUCKLAND

First published in Great Britain 1993

British Library Cataloguing in Publication Data

Thompson, Arthur
Introduction to Construction Drawing
I. Title
604.2

ISBN 0-340-56823-2

Typeset in 10 on 12pt Palatino by Wearset, Boldon, Tyne and Wear
Printed in Great Britain for Edward Arnold, a division of Hodder and Stoughton Ltd, Mill Road, Dunton Green, Sevenoaks, Kent TN13 2YA by St Edmundsbury Press Ltd, Bury St Edmunds, Suffolk.
Bound by Hunter & Foulis Ltd, Edinburgh, Scotland.

Preface

The purpose of this book is to introduce construction drawing to students with no previous knowledge of the subject. The intention is that after working through the book, the student will have acquired basic construction drawing skills, and can then proceed to further study or work experience to achieve professional competence in draughting.

The book will particularly appeal to BTEC students on full and part time construction courses. It may also be useful to those following other courses where there is a drawing element. These might include Foundation Courses in Construction, examinations of the Regional Examining Unions, and examinations of the City and Guilds of London Institute.

Construction projects rely on drawings as a means of communication and it is therefore important that construction students acquire the ability to understand and produce drawings.

The book follows a "teach yourself" format, beginning from first principles. Care has been taken to avoid too much detail, which could confuse in the early stages. The drawings are all of a very simple nature. The exercises and self assessment guides included at the end of each chapter will enable students to check that drawing skills are being acquired.

After working through all the exercises the student will have built up a useful portfolio of work, the majority of which will be on A4 sheets of paper.

There is a limited amount of deliberate repetition between some of the chapters to reinforce the student's learning of key items.

A. Thompson
1993

Acknowledgements

The author is indebted to John Clancy, Chartered Architect and Chartered Town Planner for producing all the final drawings and illustrations from the author's draft originals.

He also desires to express his gratitude to the publishers, especially to Dr. Russell Parry, Senior Editor, Science, Engineering & Technology for his help and encouragement.

Finally the author wishes to thank his wife for the help she gave to him over the period of preparation of this book, and in particular for checking the typescript.

Contents

Preface iii

Acknowledgements iv

1 INTRODUCTION 1
Purpose of construction drawing; standard method of presentation; general approach; drawing materials and equipment generally; drawing sheet sizes; drawing paper; tracing film; drawing boards and alternatives; backing sheets; paper fixings; pencils; pens; ink; erasers and erasing shields; scales (scale rulers); set-squares; compasses and circle aids; other equipment; care of equipment; keeping equipment clean; self assessment guide.

2 DRAWING LINES AND SHAPES 10
Getting started; tilting the drawing board; fixing the backing sheet; fixing the drawing paper; using the straight edge or tee-square; using the set-square; general drawing strategy; drawing construction lines generally; drawing final lines generally in pencil; drawing in ink; exercises; self assessment guide.

3 DRAWING TO SCALE 19
Introduction; the metric system; using a 1:5 scale; using a 1:50 scale; using a 1:200 scale; using a 1:100 and other scales; self assessment guide.

4 LETTERING AND DIMENSIONING 23
Introduction; quality; methods of lettering; advantages and disadvantages of freehand lettering; requirements of letters and numerals; types of letters; need for guide lines; size of letters; spacing of letters; need to practise lettering; annotation; dimensioning; horizontal dimensions; vertical dimensions; final point; exercises; self assessment guide.

5 GRAPHIC CONVENTIONS 31
Introduction; lines; section lines or planes; staircases and ramps; representation of materials; doors; windows; symbols; orientation of plans; abbreviations; exercises; self assessment guide.

6 ORTHOGRAPHIC PROJECTION 41
Introduction; plans; elevations; sections; layout of orthographic projection drawings; exercises; self assessment guide.

7 ISOMETRIC AND AXONOMETRIC PROJECTION 52
Isometric projection; axonometric projection; exercises; self assessment guide.

8 TYPES OF CONSTRUCTION DRAWINGS 63
Introduction; planning and survey drawings; architectural preliminary sketch design drawings; architectural design drawings; architectural production drawings; structural and civil engineering drawings; building engineering services drawings; sub-contractors' and suppliers'

drawings; contractors' drawings; measured drawings; record drawings; freehand drawings; self assessment guide.

9 SITE PLANS 67
Introduction; purpose; scale; dimensions; datum; landscaping; services; north point; exercises; self assessment guide.

10 LOCATION FLOOR PLANS 72
Introduction; purpose; scale; dimensions; walls and partitions; windows; doors; other items; room names and notes; north point; exercises; self assessment guide.

11 LOCATION SECTIONS 78
Introduction; purpose; scale; dimensions; walls and partitions; other structural elements; other elements; exercises; self assessment guide.

12 LOCATION ELEVATIONS 83
Introduction; purpose; scale; dimensions; depicting materials; sub titles; lettering; lines; exercises; self assessment guide.

13 ASSEMBLY DRAWINGS 90
Introduction; purpose; scale; dimensions; lines; hatching; annotation; reference points; examples of assembly drawings; exercises; self assessment guide.

14 COMPONENT DRAWINGS 99
Introduction; purpose; scale; dimensions; lines; hatching; annotation; examples of component drawings; exercises; self assessment guide.

15 STRUCTURAL ENGINEERING DRAWINGS 104
Generally—introduction; types of structural engineering work; structural detailing; standardisation. Structural steelwork—types of drawings; scales; mark references for beams and columns; fixing methods; lines; dimensions. In situ reinforced concrete structures—types of drawings; scales; mark references for columns and beams; drawing reinforcement; reinforcement identification; lines; dimensions. Exercises; self assessment guide.

16 SERVICES DRAWINGS 118
Types of building services work; main groups of drawings; use of grids; identification of services components and equipment; general layouts; plant area drawings; details of specific services items; exercises; self assessment guide.

17 FREEHAND DRAWING 126
Use of freehand drawings; drawing techniques; production information sketch drawings; exercises; self assessment guide.

18 MORE ADVANCED CONSTRUCTION DRAWING 134
Introduction; advanced architectural drawings; advanced structural engineering drawings; advanced services drawings; landscape drawings; geometrical drawing; computer aided draughting; self assessment guide.

19 ORGANISATION AND MANAGEMENT OF CONSTRUCTION DRAWING WORK 137
Decision making; drawing registers; drawing numbers; checking drawings; check lists; check list for organisation of production information; drawings checklist; self assessment guide.

20 WORDS AND PHRASES 143

21 EXERCISES AND SELF ASSESSMENT QUESTIONS 146
Additional exercises and assignments
Standard self assessment questions

Bibliography 149

Index 151

1
Introduction

Purpose of Construction Drawing

Construction drawing is a means of showing in a graphical form the shape, size and position of a building on a site, together with the composition of the materials used and the way the building is to be constructed or put together. The information on construction drawings has to be presented in a precise, unambiguous way so that it can be understood by anyone with a knowledge of draughtsmanship and construction.

It should be borne in mind that the contractor's staff using the drawings on a construction site often work under difficult circumstances, and the quality and clarity of the drawings should reflect this fact. Students should therefore always aim to produce construction drawings of a good standard, which means they will need to be industrious, methodical, accurate and tidy in their work.

In this book the term 'construction drawing' includes not only drawings produced by architects, which generally make up the majority of the drawings for a building project, but also structural drawings which are the province of structural engineers, and building engineering services drawings which are commonly prepared by building services and mechanical engineers.

Standard Method of Presentation

The aim of providing precise, unambiguous information is helped by standardizing the method of presentation. In Britain the British Standards Institution provides standards for many categories of work, including draughtsmanship. They also provide standards for components, including those used by draughtspersons in the construction industry. Relevant British Standards for workmanship, materials and components related to construction drawing are listed at the back of this book.

General Approach

A number of copies of each drawing will generally be required. The person responsible for producing the drawings (architect, structural engineers etc.) will require copies for their own use; the builders will require one or more copies, as will other people involved in the project, such as council building officials. In addition the various members of the design team will exchange drawings with each other. For example, the architect will generally receive copies of all the structural engineer's drawings and the structural engineer will be given some of the architect's drawings.

It is usual therefore for the final drawing to be prepared on tracing paper, or other transparent material, so that copies can be made easily. The commonest and cheapest method of obtaining these copies is by dye-line printing. The original drawing, produced on translucent tracing paper or film, is laid over a sheet of sensitized paper and the two sheets are passed through a printing machine and exposed to a strong ultraviolet light. The sensitized paper is then developed either by exposing it to ammonia gas or to a liquid developer.

To obtain a dye-line print with black lines on a white background it is advisable to draw the original using ink or a 'black' pencil lead—e.g. an HB. Harder pencils do not generally give such good results.

If the original drawing has been prepared on opaque paper, prints true to scale can be obtained by using a gelatine plate to recreate the original work. Drawings can also be reproduced by photographic means.

Sometimes final drawings are drawn directly on to transparent tracing paper or film, but it is suggested that you begin by drawing on plain white paper, and then when you are satisfied with this initial drawing you can trace off the final copy on to tracing paper. It is further suggested that you practise using pencil first, and progress to ink drawings later.

Drawing Materials and Equipment Generally

There is a wide choice of materials and equipment available for the draughtsperson to use, but in the early stages of gaining drawing skills it is best to limit your purchases to essential items. It is sensible to buy only good quality products, for the novice will not produce high quality work with poor quality equipment.

Drawing Sheet Sizes

Construction drawings are generally produced on standard sheets known as A sizes. These conform to the recommendations of the International Standards Organisation (ISO). The largest size sheet is the A0 followed by the A1, A2, A3 and A4 sheets. The area of each sheet is half the area of the next largest size. Thus the A0 sheet is 1 sq.m.; A1 is 0.5 sq.m.; A2 is 0.25 sq.m.; A3 is 0.125 sq.m.; and A4 is 0.0625 sq.m. Details of the sheet sizes are illustrated in Fig. 1.1.

Paper can be obtained both in pre-cut sheets in the A sizes, or in a range of standard width rolls which can be cut into A sized sheets by the draughtsperson.

Drawing Paper

The three main types of paper used for construction drawings are detail paper, cartridge paper and tracing paper.

Detail paper is cheap paper and commonly available in a range of weights from 30–75 grammes/sq.m. If the lighter weights of paper are used they are generally thin enough, and consequently translucent enough, to enable dye-line prints of a readable quality to be obtained.

Cartridge paper is of better quality and therefore more expensive. It is commonly available in weights of 110–250 grammes/sq.m. It is thicker and stronger than detail paper, and cannot be used for obtaining copies from a dye-line printer, although true to scale prints or photographic copies can be obtained if required.

Tracing paper, as its name suggests, is translucent and when placed over a drawing prepared on detail or cartridge paper enables a copy of the original drawing to be traced. Good quality copies can be obtained from a dye-line printer.

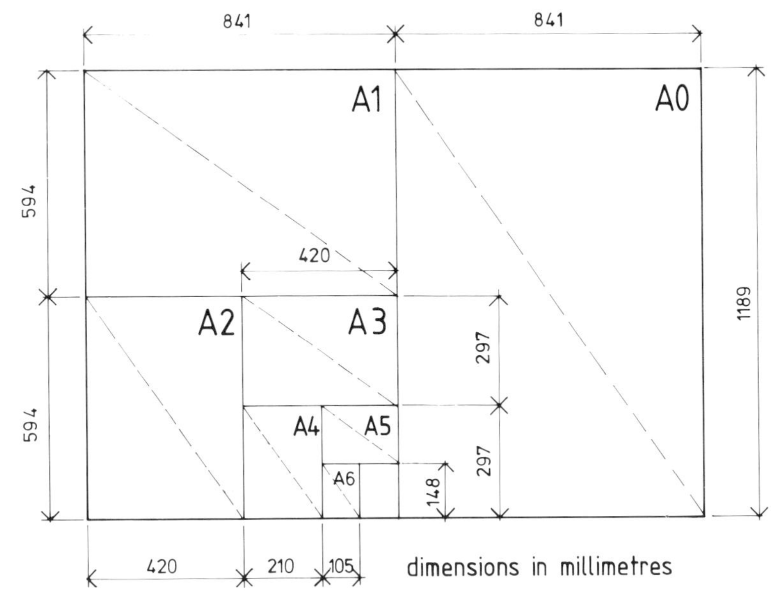

Fig. 1.1 Drawing paper standard ISO sizes

Tracing Film

Where drawings need to be kept for long periods, these may be produced on drawing film. This is a durable, high quality material which does not deteriorate with age. The two main types are acetate film and polyester film. They are more expensive than tracing paper, and unlikely to be used by the novice.

Drawing Boards and Alternatives

In order to produce a precise drawing it is necessary to fasten the paper to a sound, even and smooth surface. The traditional way of doing this is to fix it to a drawing board. This is a flat timber or plywood board, sometimes surfaced with plastic, and with its edges square and parallel. The left hand edge is generally inlaid with ebony or plastic to give a precise straight edge for the tee-square to move along (Fig. 1.2A).

The tee-square is used in conjunction with a drawing board; its purpose is to enable horizontal lines to be drawn. The tee-square consists of a horizontal straight edge or blade, 25 mm longer than the drawing board, with a vertical stock fitted at right angles to one end of the straight edge. The edge of the vertical stock is also generally finished with ebony or plastic.

The parallel motion board has largely replaced the drawing board and tee-square. A traditional type comprises a board and a straight edge for drawing horizontal lines, with an arrangement of wires, pulleys and weights which permit the straight edge to be moved easily up and down the board. The board is often mounted on a stand, with facilities for adjusting the height and tilt of the board (Fig. 1.2B).

Various other proprietary forms of parallel motion boards are available in which the horizontal straight edge is enabled to move up and down the board by guide rails in the vertical edges of the drawing board. The novice will find the parallel motion board easier to use than the drawing board and tee-square.

Professional draughtspersons may use a draughting machine. This will be mounted on a stand and will incorporate additional features to those provided on a parallel motion board. They are expensive, and the novice is advised to gain experience with a parallel motion board or drawing board and tee-square before buying one.

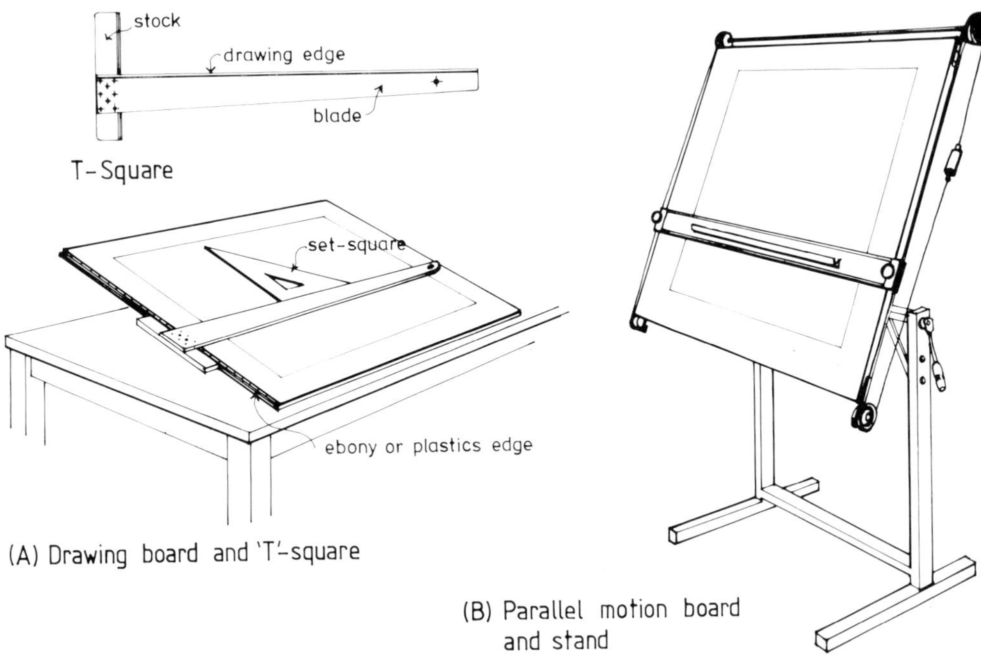

Fig. 1.2

Backing Sheets

If the finished surface of the drawing board is timber or plywood it is advisable to cover it with a backing sheet of thick cartridge paper, over which the drawing paper will be placed. This will not be necessary if the board has a melamine plastic finish unless the drawing is being prepared on tracing paper.

As an alternative to cartridge paper, proprietary backing sheets of materials such as cellulose-acetate are available.

Paper Fixings

The paper needs to be securely fixed to the drawing board. The traditional way of achieving this is by drawing pins. This method has the disadvantage of gradually causing the corner surfaces of the board to be damaged. In any case it is not a feasible method if the finished surface is melamine plastic.

Steel spring clips can be used for fixing, if the paper is the same size as the drawing board. (Fig. 1.3A).

Drafting tape is commonly used. (Fig. 1.3B).

Pencils

The student should begin by practising drawing using pencils. The traditional wooden shelled pencil has a central core of graphite and clay known as the lead, approximately 2 mm diameter, and an outer case of cedarwood.

The hardness of the lead is governed by the proportions of graphite and clay present. The more clay in the lead the harder it will be. Pencils are graded according to their hardness. 9H is the hardest available pencil, and 8B the softest. The letter H denotes hardness and the letter B blackness. An HB pencil is in the centre of the scale in respect of both hardness and blackness. The harder the pencil lead is the less likely the drawing is to smudge, but the less black the line will be. An HB lead is probably the best lead for the beginner when preparing architectural

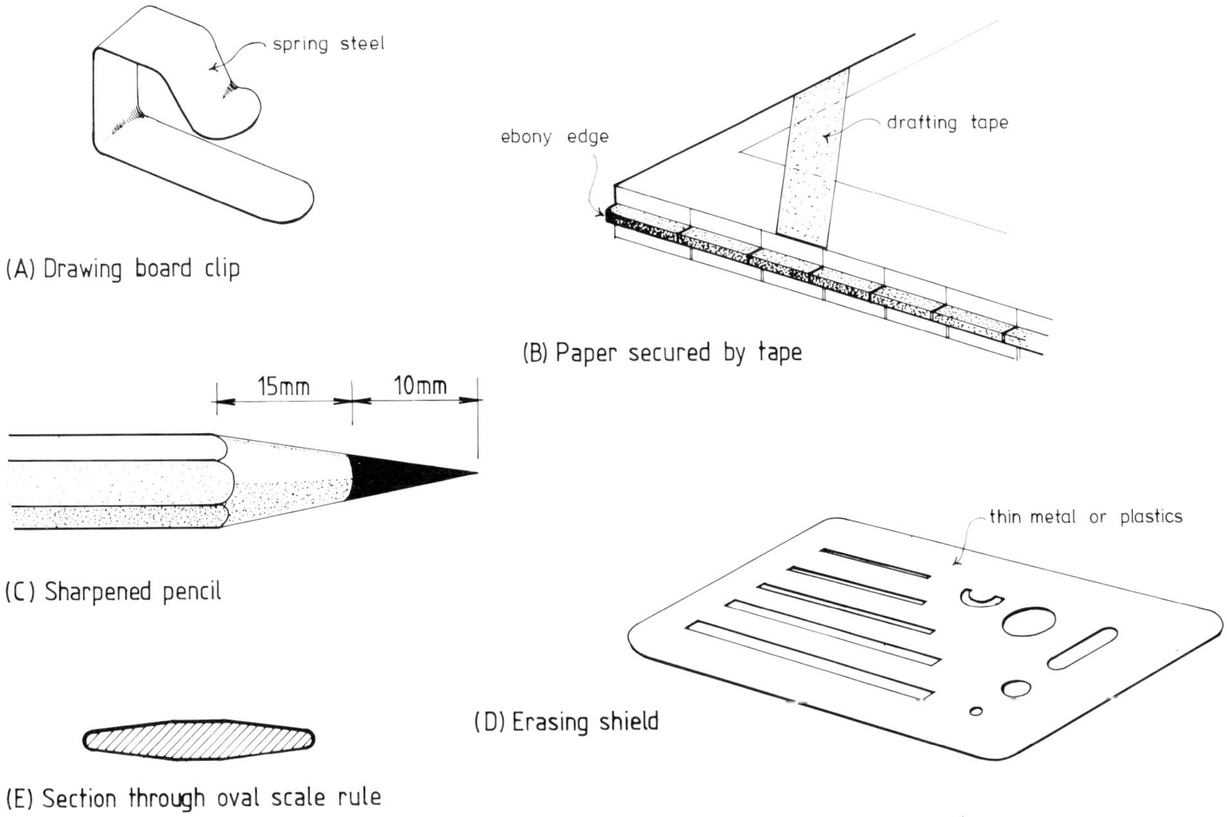

Fig. 1.3 Drawing aids

drawings, although some draughtspersons use an H or 2H for drawing construction (guide) lines and an H or HB for the final (finished) lines. Engineers generally use pencils with harder leads. Degrees of lead hardness are not necessarily consistent between different pencil manufacturers. There is an advantage therefore in always using pencils from the same manufacturer, and it is certainly worthwhile to use only good quality pencils.

The traditional pencil can be sharpened by a pencil sharpener with a cutting blade. Alternatively a penknife can be used to sharpen it to a conical point, with the exposed lead about 10 mm long, as shown in Fig. 1.3C.

An alternative to the traditional pencil is the mechanical clutch pencil. This consists of a metal holder into which 2 mm diameter leads are inserted and secured by a clutch. Instructions for their use will be supplied by the manufacturer, but generally all you have to do to prepare them for use is to push the button at the top end of the pencil to adjust the lead. The button is often removable and serves as a lead sharpener. If this is not the case the lead can be sharpened to a conical point using a sandpaper block.

Another alternative is the fine-lead mechanical pencil, which uses slender leads which are protected by a steel sleeve. The lead diameters exactly relate to the line thicknesses. I.e. a 0.25 mm lead draws an 0.25 mm thick line; an 0.5 mm lead draws an 0.5 mm line, and so on. When using fine-lead mechanical pencils care must be taken not to use too much pressure, or the leads will break.

The choice of drawing equipment is to a large extent a matter of personal preference. The author prefers the fine-lead pencil, with the traditional pencil as his second choice, and has found that many students share this preference.

Pens

Many different makes of fountain pen are available for drawing in ink. They are given various names, such as technical pens and tubular point pens.

Pens are designed to give a variety of line thickness. It is suggested that the beginner can manage with two pens initially: a thinner pen for general work (0.25 mm will be suitable) and a thicker pen for lines needing to be emphasized (0.5 mm will be suitable.)

The pens are supplied by the manufacturer with a clear set of instructions on how to use them, and it is important to follow these instructions closely, particularly with regard to the procedure for keeping the pens clean.

Ink

The safest procedure with regard to inks is to use the type recommended by the pen manufacturer. Unless the pen manufacturer gives contrary advice, a suitable drawing ink will be one which is waterproof and smudgeproof when dry, but can be erased.

Erasers and Erasing Shields

A variety of soft erasers are available which will easily erase pencil lines without damaging the surface of the paper. It is best to avoid large erasers as they tend to harden and deteriorate with age.

Solvent erasers can be used to erase ink work from tracing paper, but hard erasers are needed to remove ink work from cartridge paper.

Erasing shields, which are thin pieces of metal or plastic with different shaped openings formed in them, (see Fig. 1.3D), are a useful aid to the erasing process. These shields protect the areas of drawing close to the work which has to be erased.

Scales (Scale Rulers)

Drawings are drawn in proportion to the actual size of the building. This is known as the scale of the drawing. If you use a scale of 1:50 the object you are drawing will actually be 50 times large than your drawing. This matter is discussed in Chapter 3.

Scale drawings are generally produced with the aid of scale rulers, which are commonly referred to as scales. They are narrow strips of plastic or hardwood, and either flat, oval (see Fig. 1.3E) or triangular in section, with divisions marked along each edge and all sides of the scale. They are generally either about 300 mm or 150 mm long.

An oval section scale that tapers to a thin edge, so that it can be brought close to the paper to enable the dimensions to be read off, is probably easiest for the beginner to use. It is suggested you buy a scale graduated in the following proportions—1:10, 1:100, 1:20, 1:200, 1:5, 1:50, 1:1250, 1:2500. In any event make sure that 1:5, 1:50, 1:100 and 1:200 scales are included.

Set-Squares

Set-squares are used in conjunction with the horizontal straight edge of a parallel motion board or a tee-square to draw vertical and inclined lines. They are commonly made of clear plastic, and may have square or bevelled edges.

Fixed set-squares generally have angles at either 45 degrees or 30/60 degrees (Fig. 2.1). It is suggested that the student buys one of each type. Set-squares with square edges, with the longest side of about 230 mm will be suitable.

Adjustable set-squares are really a set-square cut in two, pinned at one point and adjusted on a clamp which allows different angles to be

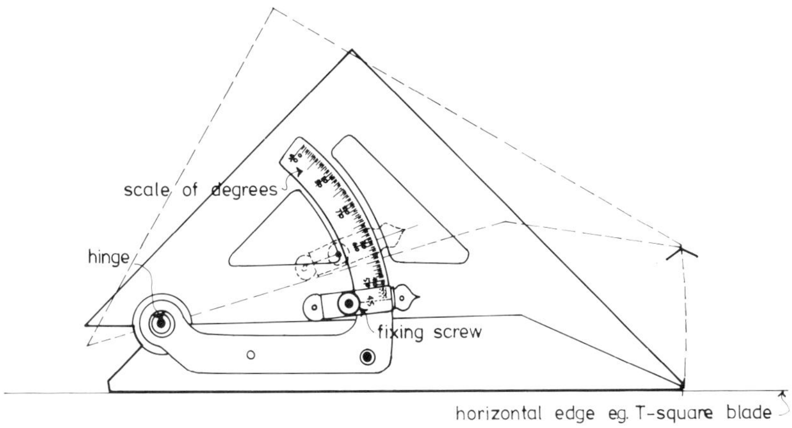

(A) Adjustable set-square
broken line shows square adjusted to draw angles 60° and 30°

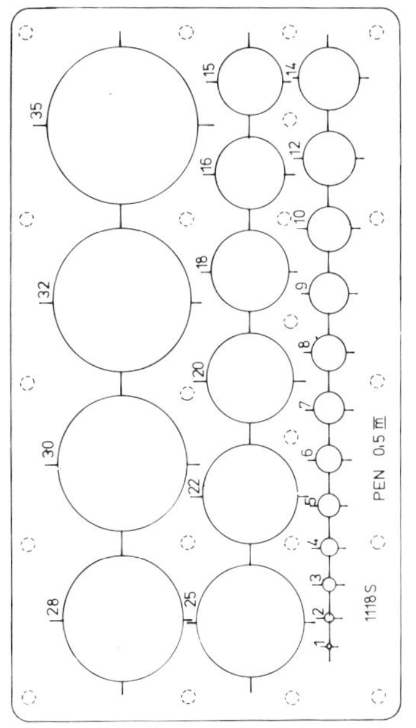

(B) Circle template

Fig. 1.4

drawn (see Fig. 1.4A). They are expensive and students should be able to manage without an adjustable set-square in the early stages of their careers.

Compasses and Circle Aids

Lines on construction drawings are mainly straight, but sometimes it will be necessary to draw circular lines. Circular lines are traditionally drawn using compasses with various attachments, including those supplied by manufacturers of technical pens to enable their pens to be used in conjunction with a compass.

Special circle templates can be used instead. A plastic circle template is shown in Fig. 1.4B. If students buy one of these they should be able to dispense with compasses initially.

Other Equipment

In addition to the main items of equipment already mentioned, there are various other items which are available to aid draughtspersons in their work.

Protractors are used for setting out angles. They are commonly semi-circular, manufactured in clear plastic, and divided into degrees. You will occasionally need these if you do not possess an adjustable set-square.

Clear plastic French curves are available to draw irregular curves. A wide variety of templates, apart from circle templates, are also available in clear plastic.

Dividers may be used for transferring distances from one drawing to another. Beam compasses are used for drawing large circles. Electric erasing machines are available.

Care of Equipment

If students buy good quality equipment, it will give long and satisfactory service provided it is treated with care, and not misused.

Drawing boards should never be stood on their ebony or plastic edges.

Tee-squares and the straight edges of parallel motion boards should never be used as a guide for a cutting knife. Once nicked the tee-square or straight edge becomes useless for drawing smooth, even lines. When the tee-square is not in use, it should rest on the drawing board, making sure that the whole of the blade is fully supported.

Set-squares should be used with similar care, as a single nick in the edge of a set-square will ruin it for drawing work.

Scales should only be used as scales, never as letter openers, or as a straight edge for ruling lines or cutting paper.

Keeping Equipment Clean

If drawings are to be kept clean it is obviously important to keep all equipment, as well as hands, clean too. Manufacturers recommendations regarding the best methods of cleaning their products should be followed in all cases. If in doubt, soap and water is generally safe and effective. The top surfaces of boards can be brushed with a soft dust brush to remove erasing crumbs etc. Tee-squares can be cleaned using a cloth slightly dampened with turpentine substitute.

It is particularly important to keep set-squares clean if dirty drawings are to be avoided.

Care should be taken to keep technical pens clean, as otherwise they will become clogged.

Self Assessment Guide

In order to check whether you have understood all you have read in this chapter, you should now attempt to answer the following questions.

1 What is the purpose of construction drawing?
2 After a drawing has been prepared several copies are generally required. Why is this?
3 Is an A0 sheet of drawing paper larger or smaller than a A1 sheet?
4 When a tee-square is in position on a drawing board it has two parts—the horizontal part and the vertical part. What is the name given to the horizontal part?
5 What is the name given to the vertical part of a tee-square?
6 What is the disadvantage of fixing drawing paper to drawing boards with drawing pins?
7 Out of 'H' 'HB' and '2H' pencil leads, which is the hardest lead, and which is the softest lead?
8 What is a suitable pencil lead to use when preparing an architectural drawing?
9 When preparing their drawings do engineers generally use harder or softer pencil leads than architects?
10 Which type of pencil does not require sharpening?
11 It was suggested in Chapter 1 that initially the beginner should use two pens. What were the suggested line thicknesses of these two pens?
12 If you are using a 1:50 scale to draw a plan of a building, how many times bigger will the actual building be than your plan?
13 What is the purpose of a set-square?
14 What is the difference between a fixed set-square and an adjustable set-square?
15 How can circles be drawn other than by using a compass?
16 What is the purpose of a protractor?
17 Why must a tee-square never be used as a straight edge for a cutting knife?
18 Why is it important to keep technical pens clean?
19 Why should all drawing equipment be kept clean?

When you have attempted all the above questions, reread the appropriate part of the chapter and check you know the correct answer to every question.

2

Drawing Lines and Shapes

Getting Started

You should now have a general idea as to what construction drawing is all about, and the equipment available to assist the draughtsperson.

At this stage you should have the following equipment.

- Sheets of A4 and A3 cartridge or detail paper, and tracing paper.
- Preferably an A2 parallel motion board. Alternatively an A2 drawing board and tee-square may be used. If you are left handed it may be easier for you to work on a parallel motion board, although left handed tee-squares are available.
- Drafting tape. If this is not available, sellotape can be used instead.
- Fine-lead mechanical pencils with H and HB leads. Two thicknesses of lead are required: the thinner should be 0.2–0.25 mm, and the thicker lead 0.4–0.5 mm. As an alternative to fine-lead mechanical pencils traditional wooden shelled pencils or mechanical clutch pencils may be used instead.
- Two technical pens, one for drawing lines 0.2–0.25 mm thick and the other for drawing lines 0.4–0.5 mm thick.
- Ink of a type recommended by the technical pen manufacturer.
- A soft eraser and a solvent eraser or a hard eraser.
- A 300 mm long oval scale which includes scales of 1:100, 1:200, 1:5 and 1:50.
- A 45 degree fixed set-square and a 30-and-60 degree fixed set-square—longest side to be about 230 mm.
- A plastic circle template.
- At a later stage a protractor or adjustable set-square will be required.

The intention of this chapter is to give you practice in drawing lines and shapes in pencil and ink, on cartridge/detail paper and tracing paper.

You will use two line thicknesses. The thinner lines should be 0.2–0.25 mm thick and the thicker lines 0.4–0.5 mm.

Tilting the Drawing Board

The drawing board should be slightly tilted towards the draughtsperson. Most parallel motion boards and some drawing boards are designed to achieve this. In other cases triangular blocks of wood, or bricks, can be used to produce the tilt.

Fixing the Backing Sheet

A backing sheet should be fixed to the board with drafting tape. The sheet should be exactly the same size as the drawing board. Make sure that the top edge of the paper is exactly in line with the top edge of the board.

Fix the top left hand corner of the backing sheet first. Then fix the bottom right hand corner, pulling the paper tight, without stretching it. Then fix the top right hand corner, followed by the bottom left hand corner.

Fixing the Drawing Paper

Now fix a sheet of white A3 paper to the board with drafting tape in a similar fashion to that used for fixing the backing sheet. The sheet should be placed with the long side of the paper exactly horizontal—i.e. the top (long) edge of the paper should be exactly parallel to the top edge of the board. It is best if the edges of the paper are a minimum of 30 mm away from the edges of the board. Use your straight edge (if you are using a parallel motion board) or your tee-square (if you are using a drawing board) to ensure that the paper is fixed horizontally.

Using the Straight Edge or Tee-Square

If you are using a parallel motion board move the straight edge carefully up or down to the required position. It is best to hold the straight edge as near to its centre as possible. Do not actually draw any lines at this stage, but just get used to moving the straight edge up and down and holding it gently but firmly in position.

If you are using a tee-square, press the stock of the tee-square against the left hand edge of the board and allow it to slide up and down until the blade is in the required position. Do not actually draw any lines at this stage. When you do so, you will have to hold the tee-square firmly with the left hand while drawing lines with the right hand. The pencil should be held against the ruling edge of the tee-square blade.

Using the Set-Square

You will need to use your set-square for drawing vertical and sloping lines. Move the straight edge/tee-square to the required position. Place the set-square on the straight edge/tee-square with its base on the top edge of the blade, and the vertical edge in the required position (see Fig. 2.1). Hold both the straight edge/tee-square and set-square gently but firmly in position with the left hand, while leaving the right hand free to draw the line. Do not actually draw any lines yet, but just become familiar with moving the straight edge/tee-square and set-square around and holding them in position.

General Drawing Strategy

It is generally best to start a drawing from the top left hand corner of the paper, and work across the sheet and downwards towards the bottom right hand corner. This helps to avoid going over lines previously drawn and possibly smudging them. A left handed person can start at the top right hand corner and work downwards towards the bottom left hand corner.

A drawing may contain a single view, or a number of views—e.g. plans and elevations—on the same sheet. In cases where there are several views included on the one drawing sheet, the usual procedure is to work up all the views at the same time, rather than completing one view before starting the next one. Reference lines should be drawn first, followed by the main outline of each view, and then the detail can be added.

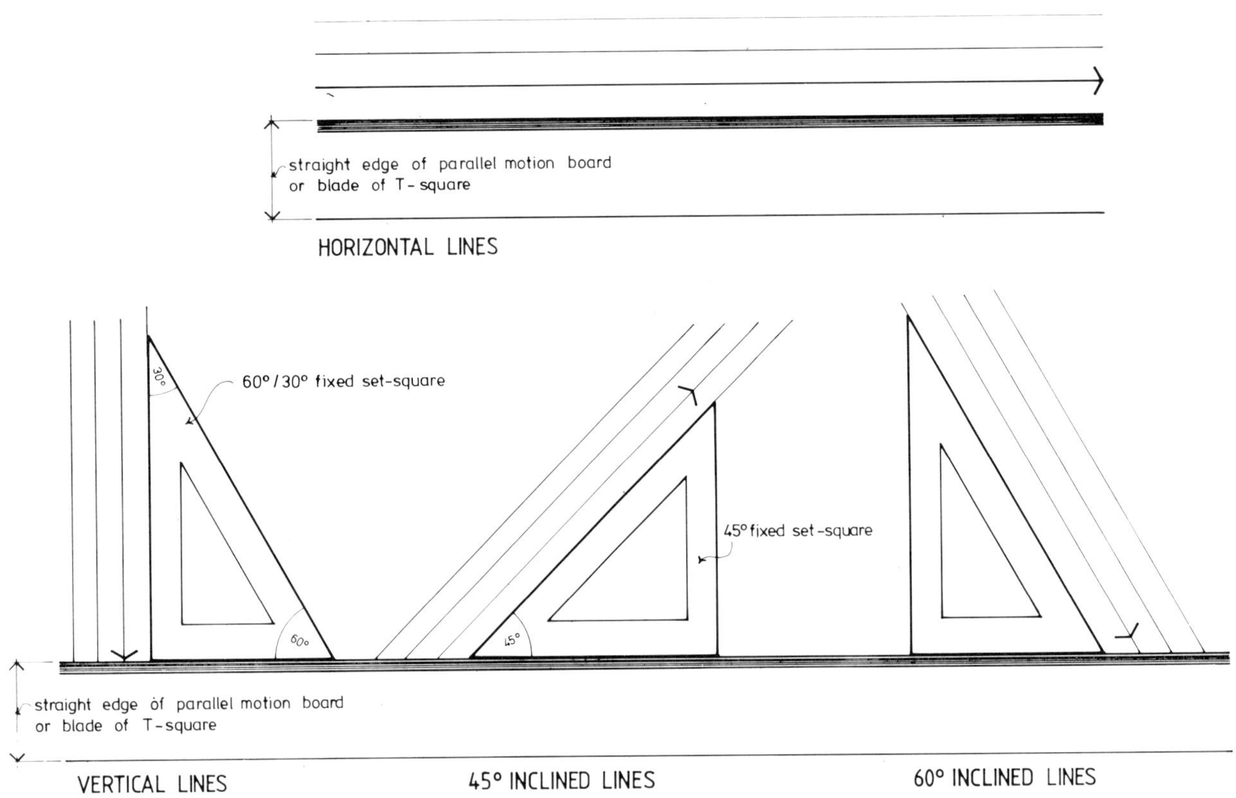

Fig. 2.1 Drawing straight lines

Care should be taken to arrange the views on the sheet so that there is room for notes and dimensions, with sufficient space left around the views to prevent the drawing looking crowded.

Drawing Construction Lines Generally

All lines should be drawn very lightly first in H or HB pencil, so that they are just visible to the draughtsperson. This is achieved by applying the minimum pressure to the pencil and moving it very gently over the paper. The lines produced are guide lines, generally called *construction lines*, and they are drawn longer than the finished (or final) lines. When the construction lines have been completed, the drawing can be lined in with clear, firm final lines, generally with an HB pencil.

Drawing Final Lines in Pencil

Final lines should be firm, black, of even quality, and of consistent thickness throughout their length.

In the case of the thinner lines, regardless which type of pencil is used, the lines should be produced by a single stroke of the HB pencil.

In the case of the thicker lines, when using a traditional wooden shelled pencil or a mechanical clutch pencil, a second line needs to be drawn alongside the original line to increase it to the required thickness. Be sure to hold the straight edge/tee-square and set-square firmly in position with one hand while you draw the pencil along the ruling edge with the other hand. Incline the pencil slightly, keeping the angles with the straight edge/tee-square constant, and rotate it continuously to maintain the point of the pencil.

Drawing in Ink

When drawing in ink the instructions provided by the manufacturer of the technical pens should be closely followed, but generally pens should be held almost upright against the ruling edge of the straight edge/tee-square or set-square and drawn smoothly across the paper, using light, even pressure.

When final ink drawings need to be prepared on cartridge paper it will be necessary to draw faint pencil construction lines first, and then line them over with ink afterwards. It is important to ensure that the initial pencil construction lines are completely blotted out by the subsequent ink lines.

It is more usual to prepare ink drawings on tracing paper, and in this case the best practice is to complete the drawing in pencil on detail paper first, and then to lay a sheet of tracing paper over the detail paper and trace the drawing in ink.

Exercise 2.1

Drawing horizontal, vertical and inclined lines

This exercise is intended to give you practice in drawing horizontal, vertical and inclined lines in pencil.

All construction lines should be drawn first, followed by all final lines.

The completed exercise should look as illustrated in Fig. 2.3.

Drawing the border and title panel

Now that the drawing paper is fixed to the board and you have some general information about drawing lines, the first operation is to draw the border and title panel.

The border should be drawn around the four edges of the paper 10 mm wide. Initially just draw the construction lines for the border.

Title blocks are required on drawings. These are generally either arranged as a continuous horizontal strip along the bottom of the drawing; a continuous vertical strip up the right hand side of the drawing; or a block in the bottom right hand corner of the drawing. The work shown in this book uses a continuous horizontal title strip.

Form the title block by drawing a construction line 20 mm up from the bottom border line. Add the short vertical and horizontal lines so that the title panel looks as illustrated in Fig. 2.2.

Now divide the area inside the border into four equal rectangles by drawing one horizontal construction line and one vertical construction line. Your drawing should now look as illustrated in Fig. 2.2.

You are now ready to practise drawing straight horizontal, vertical and inclined lines. All the lines should be drawn as construction lines first, and then converted into final lines afterwards.

Drawing horizontal construction lines

The horizontal construction lines should be drawn first, in the top left hand rectangle on the drawing sheet. The lines should extend from near the left hand border line to near the vertical central division line.

By working from the top to the bottom of the sheet you will move the straight edge/tee-square away from the line just drawn instead of rubbing over it.

The lines should be spaced about 10 mm apart. Start with the top line, moving the pencil from the left to the right (see Fig. 2.1). Apply the pencil lightly to the paper. One pass only is required for each line, and they should be faint and just visible.

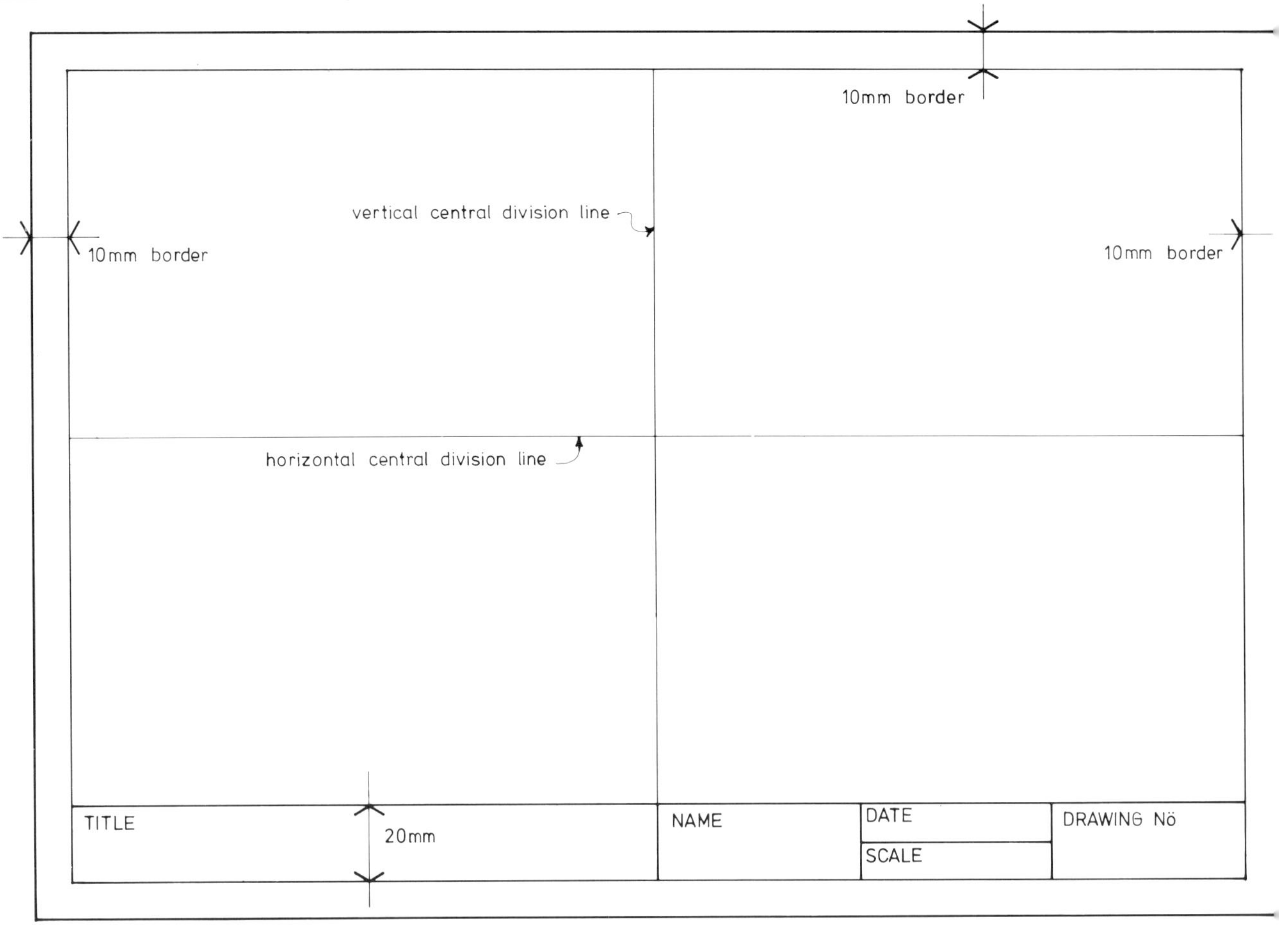

Fig. 2.2 Preparing drawing sheet for Exercise 2.1

Drawing vertical construction lines

The vertical construction lines should be drawn next in the top right hand rectangle of the drawing sheet. The lines should extend from near the top border line to near the horizontal central division line.

The lines should be spaced at about 10 mm apart. Start with the left hand line moving the pencil from the top to the bottom (see Fig. 2.1). Remember to use the minimum pressure on the pencil to produce faint construction lines.

To draw the vertical lines you will need to use the 60-and-30 degrees set-square. Rest the shorter, horizontal edge of the set-square on the straight edge/tee-square blade and hold it in place with the left hand, which should also hold the straight edge/tee-square in position. This will leave the right hand free to draw the lines along the vertical edge of the set-square.

Drawing inclined construction lines

The inclined construction lines should be drawn next in the bottom left hand rectangle of the drawing sheet, again occupying virtually the complete space between the borders and the central division lines. They are to be at an angle of 45 degrees, about 10 mm apart, and to run in the direction shown in Fig. 2.3.

Finally draw the inclined construction lines in the bottom right hand rectangle. The lines should run in the direction shown in Fig. 2.3.

To draw the inclined lines you will need to use the 45 degree set-square. Rest one edge of the set-square on the straight edge/tee-square blade. It

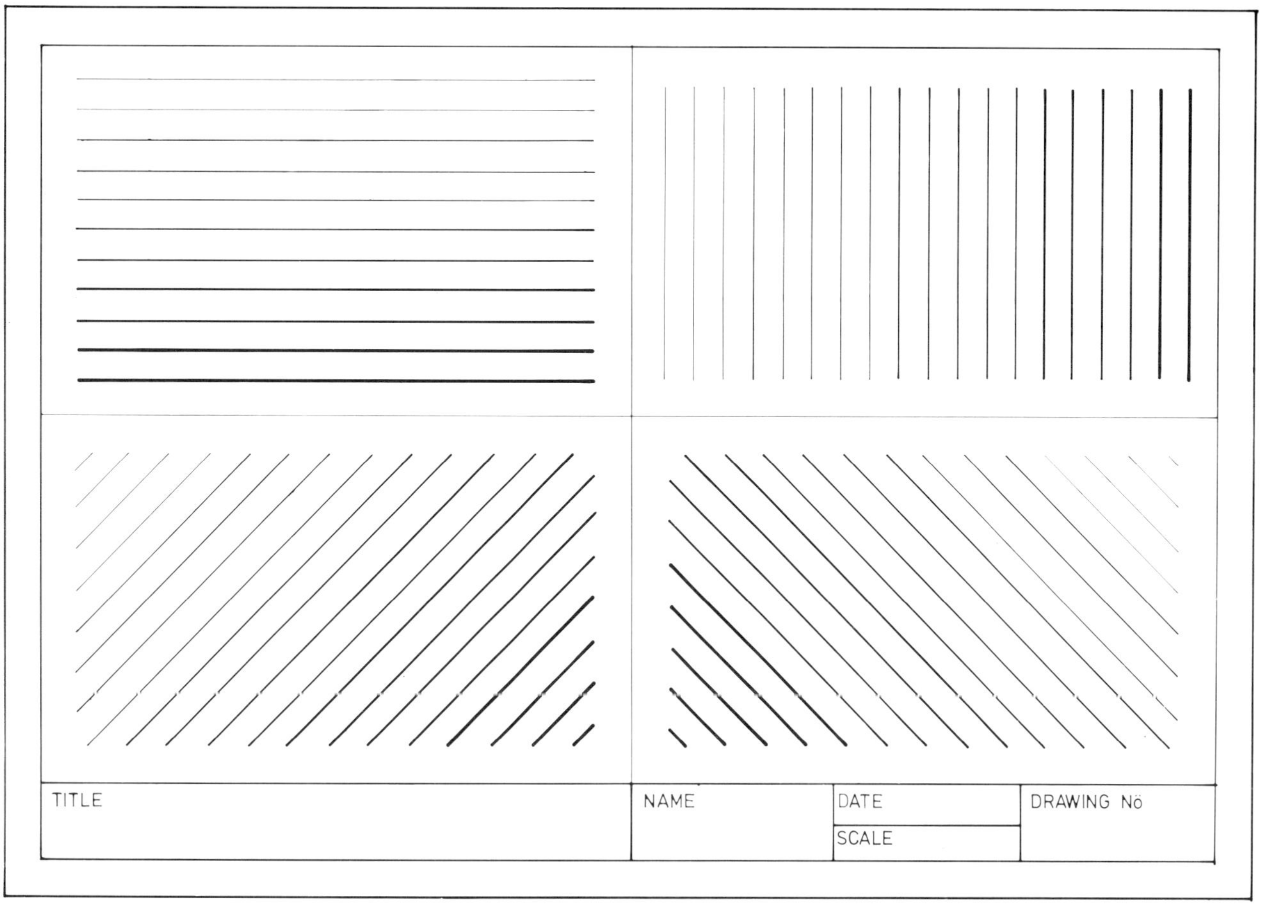

Fig. 2.3 Exercise 2.1

should be kept in place with the left hand, which will also hold the straight edge/tee-square in position. This will leave the right hand free to draw the lines along the inclined edge of the set square (see Fig. 2.1).

Drawing final lines

Now that you have drawn all the construction lines you are ready to line in the final lines. The border and title panel lines should be thicker lines—i.e. 0.4–0.5 mm. About half of the horizontal, vertical and inclined lines should be thin lines—i.e. 0.2–0.25 mm—with the remaining lines drawn as thicker ones. All horizontal, vertical and inclined final lines should stop about 10 mm short of the border. The completed drawing should look like Fig. 2.3.

As far as possible you should start drawing from the top left hand corner, and work across the sheet and downwards towards the bottom right hand corner. Remember that it is important to make all final lines dense, smooth and of consistent thickness.

Complete the title block, writing in capital letters the words TITLE, NAME, DATE, SCALE and DRAWING, exactly as shown in Fig. 2.3. Then neatly print the following information under the appropriate heading, again using capital letters.

TITLE	Horizontal, vertical and inclined lines
NAME	Your own name
DATE	The date you complete the drawing
SCALE	Leave blank
DRAWING NO.	Exercise 2.1.

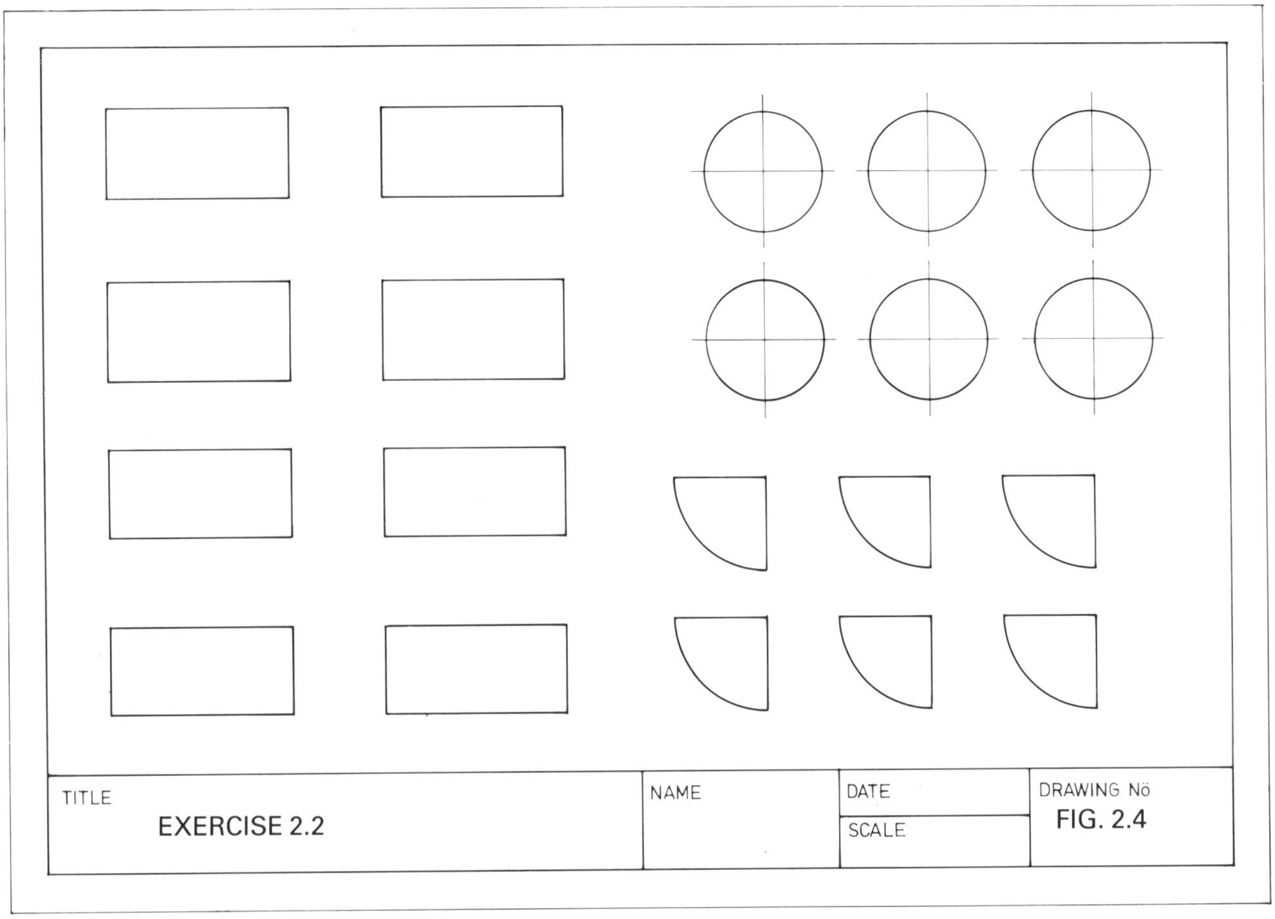

Exercise 2.2 **Drawing rectangles, circles and quadrants**

This exercise is intended to give you practice in drawing rectangles, circles and quadrants in pencil.

As with Exercise 2.1 all construction lines should be drawn first, followed by all final lines.

The completed exercise should look as illustrated in Fig. 2.4.

Drawing the border and the title panel
Draw the border and title panel in the same way described for Exercise 2.1.

Drawing rectangles
Begin by drawing the construction lines for the eight rectangles on the left hand half of the sheet. The exact sizes and shapes are not important. For each rectangle first draw two horizontal lines about 30 mm apart and about 75 mm long. Join the ends of the horizontal lines by drawing two vertical lines about 60 mm apart forming a rectangle 60 × 30 mm (see Fig. 2.5A).

When the construction lines for the entire drawing have been completed you can return to the part of the drawing containing the rectangles, and line them in. This is achieved by going over the faint construction lines, placing more pressure on the pencil as you draw over the lines defining the rectangles and making them blacker.

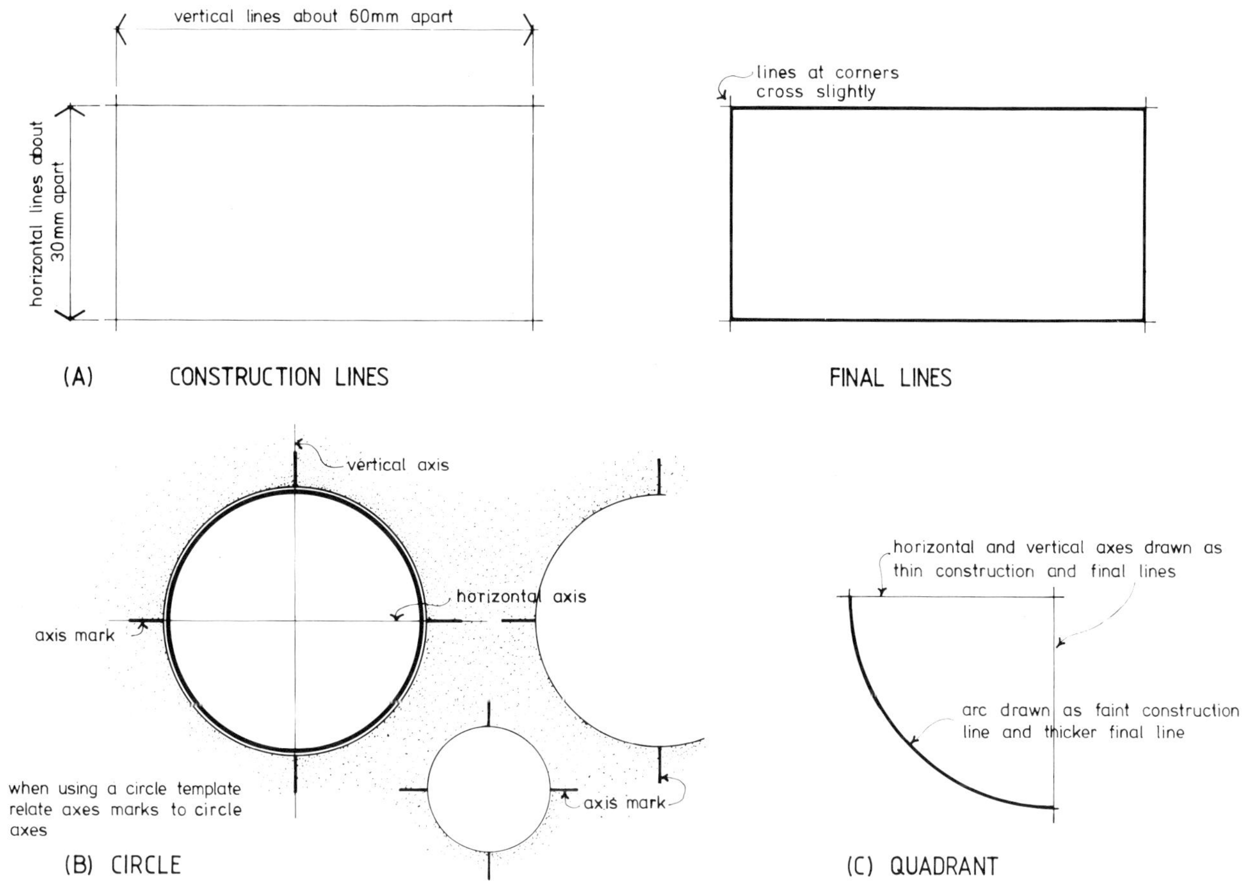

Fig. 2.5 Exercise 2.2 details

The final lines should be 0.4–0.5 mm thick. The vertical and horizontal lines should either exactly meet at the corners, or cross slightly as shown on Fig. 2.5A.

Drawing circles

After drawing the construction lines for the rectangles, draw the construction lines for the six circles in the top right hand corner of the drawing. The exact sizes are not important. First draw the horizontal and vertical axes lines for each circle. Then draw the circle. If you are using a compass to draw the circles, the compass point should be carefully placed on the precise spot where the horizontal and vertical axes cross. Then draw the circle as a curved construction line (see Fig. 2.5B). If you are using a circle template, you will need to relate the axes marks on the template with the axes drawn on your drawing sheet.

When all the work on the rectangles has been completed you can return to the part of the drawing containing the circles and line them in with final lines 0.4–0.5 mm. thick. Again draw the final line over the construction line with care, making the original line blacker by exerting more pressure on the pencil, but ensuring that double lines are not produced.

Drawing quadrants

After drawing the construction lines for the circles, draw the construction lines for the six quadrants in the bottom right hand corner of the drawing. First draw the horizontal and vertical axes of each arc. Then draw the arcs. If you are using a compass to draw the arcs, the compass point should be carefully placed on the precise spot where the horizontal and vertical axes meet. If you are using a circle template, you will need to relate the axes marks on the template with the axes drawn on your drawing sheet. Complete the quadrants by drawing the horizontal and vertical straight lines.

When all the work on the circles has been completed, you can return to the part of the drawing containing the quadrants, and line them in with final lines. Straight lines should be drawn as thin lines, and curved lines as thicker lines.

The completed exercise should look as illustrated in Fig. 2.4.

Exercise 2.3

Practising tracing in pencil

Trace in pencil, on sheets of A3 tracing paper, the whole of Exercises 2.1 and 2.2.

Exercise 2.4

Practising tracing in ink

Trace in ink, on sheets of A3 tracing paper, the whole of Exercises 2.1 and 2.2.

Exercise 2.5

Drawing on cartridge paper

Draw in ink, on a sheet of A3 cartridge paper, the whole of Exercise 2.2. N.B. Construction lines are to be drawn in pencil first, and then lined over with the final ink lines.

Self Assessment Guide

You should now check that you have gained the necessary skills in the use of the various instruments.

You should be able to move the parallel motion straight edge or tee-square up and down the board with one hand, and leave the other hand free for drawing. If you cannot, continue to practise until you can do so.

You should be able to hold the straight edge/tee-square and the set-square in any position with one hand, and leave the other hand free for drawing. If you cannot, continue to practise until you can do so.

Now look at the horizontal, vertical and inclined lines you have drawn in pencil. Are all the lines smooth, even and of constant thickness? If not, draw some more lines to try and improve your standard.

Now look at the rectangles you have drawn in pencil. Are all the lines smooth, even and of constant thickness? Do the horizontal and vertical lines either exactly meet or slightly cross at the corners of the rectangles? If not, draw some more rectangles to try and improve your standard.

Now look at the circles and quadrants you have drawn in pencil. Are all the lines smooth, even and of constant thickness? If not, draw some more circles and quadrants and try to improve your standard.

Now look at the horizontal, vertical and inclined lines, rectangles, circles and quadrants that you have drawn in ink and assess your work in the same way as you did for your pencil work. Where necessary carry out additional practice work in ink.

3

Drawing to Scale

Introduction

It is not generally feasible to draw buildings, or parts of buildings, to their actual size. Instead they are drawn in proportion to the actual measurement of the object. This proportion is known as the scale of the drawing. Common scales are 1:1, 1:5, 1:10, 1:20, 1:50, 1:100, 1:200, 1:500, 1:1000, 1:1250, 1:2500, 1:10 000. If the scale is 1:5, the object is drawn a fifth of its actual size; in other words the object is five times larger than shown on the drawing. If the drawing is 1:10, the object is drawn a tenth of its actual size, and so on.

The reader was recommended in Chapter 1 to buy an oval scale. It will probably have eight scales marked on it, i.e. two lines of divisions along each edge of both sides of the scale. Make sure it includes 1:5, 1:50, 1:100 and 1:200 scales.

It is important to state on every drawing which scale has been used. It is equally important to draw each line accurately to scale.

The Metric System

The British construction industry has adopted the metric system of measurement, which includes millimetres and metre units. There are 1000 millimetres to 1 metre. Commonly on drawings the only units used are millimetres which are abbreviated to mm. In practice the symbol 'mm' is omitted, and only the number is written. If a length on a drawing is written as 1000, this means it is 1000 mm or 1 metre, which can be abbreviated to 1 m.

Here are a few other examples. 2000 means 2000 mm or 2 m; 30 000 means 30 000 mm or 30 m; 500 means 500 mm or 0.5 m; 1600 means 1600 mm or 1.6 m; 60 means 60 mm or 0.06 m; 5 means 5 mm or 0.005 m.

It is important to get into the habit of thinking in millimetres. A few measurement examples are given below. A normal wooden shelled pencil is about 8 mm thick; an oval scale is about 30 mm wide; a foolscap envelope is about 230 mm long; the pages in this book are 297 mm high; the seat of a dining room chair is about 460 mm above the floor; a dining room table is about 740 mm high; a room in a modern home is commonly about 2300 mm high; a London taxi is about 4420 mm long; a tennis court is about 23 800 mm long.

As you can see, sometimes the thousands of millimetres are separated from the rest of the number by a slight gap. E.g. 23800 as 23 800. In effect this separates the metres from the millimetres: 23 800 is 23 m, 800 mm. Some draughtspersons use a comma for the same effect. E.g. 23,800. Others use a decimal point. E.g. 23.800. It is suggested that students do not follow these practices, and in this book numbers are written as 23800 etc.

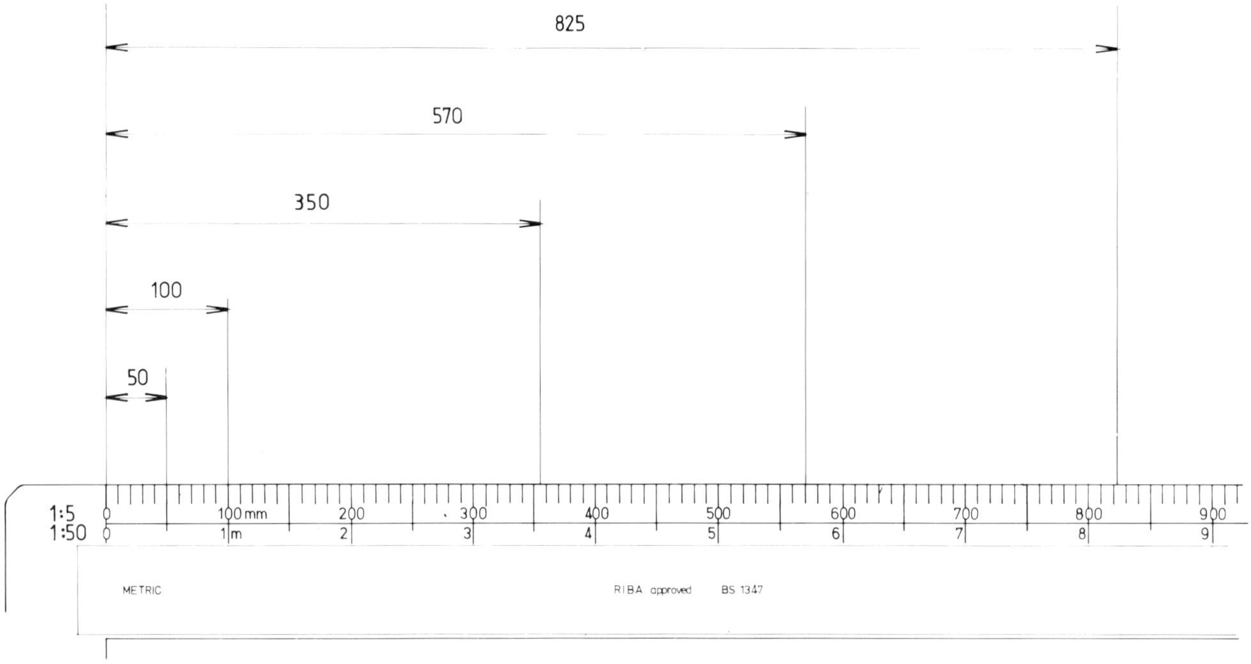

Fig. 3.1 Using a 1:5 scale

Using a 1:5 Scale

A 1:5 scale is used for drawing details of a building. An example is the eaves detail shown on drawing no. D13/4 (page 95).

Part of a 1:5 scale is shown on Fig. 3.1. It is located on the top edge of the scale. The numbering on this 1:5 scale is 0, 100 mm, 200, 300, 400, 500, 600, 700, 800, 900. Each small division on the scale represents 10 mm. Five small divisions represent 50 mm. This 50 dimension is marked on Fig. 3.1. Ten small divisions represent 100 mm; thirty five small divisions represent 350 mm; fifty seven small divisions represent 570 mm; and eighty two and a half divisions represent 825 mm. All of these dimensions are also marked on Fig. 3.1.

Now check the following dimensions on drawing no. D13/4 (see page 95).

- Plaster wall finish—14 mm. This will be represented on the scale by one small division, plus one-third of a small division.
- Wall plate width—100 mm. This will be represented on the scale by ten small divisions.
- Soffit board width—150 mm. This will be represented on the scale by fifteen small divisions.
- Wall thickness—300 mm. This will be represented on the scale by thirty small divisions.

Using a 1:50 Scale

A 1:50 scale can be used for the floor plan of a building. An example is the bungalow floor plan shown on drawing no. D10/1 (see page 75).

Part of a 1:50 scale is shown on Fig. 3.2. It is located near the same top edge of the scale as the 1:5 scale, but just below it. The numbering on this 1:50 scale is 0, 1 m, 2, 3, 4, 5, 6, 7, 8, 9. These represent metres—i.e. 1 metre, 2 metres, 3 metres and so on. However, as has already been mentioned, dimensions on construction drawings are written in millimetres, so when you read along the large divisions of the 1:50 scale, instead of reading 1 m, 2 m, 3 m and so on, you should count them off as 1000 mm, 2000 mm, 3000 mm etc.

Because there are 1000 millimetres in a metre and the large divisions are

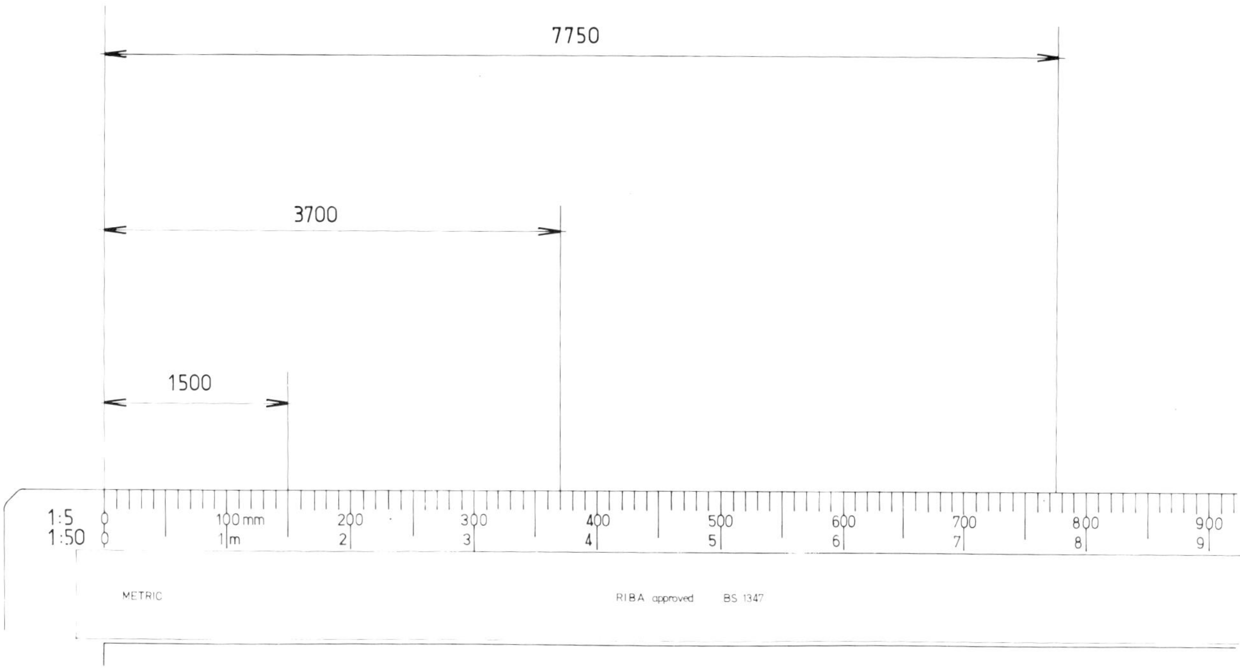

Fig. 3.2 Using a 1:50 scale

subdivided into ten smaller divisions, each of the small divisions represents 100 mm. Fifteen small divisions represents 1500 mm on the 1:50 scale. This dimension is marked on Fig. 3.2. Similarly, thirty seven small divisions represents 3700 mm and seventy seven and a half small divisions represents 7750 mm. These two dimensions are also marked on Fig. 3.2.

Using a 1:200 Scale

A 1:200 scale can be used for the floor plan of a large building, or the site plan of a small building project. An example of a site plan is shown on drawing no. D9/1 (see page 69).

Part of a 1:200 scale is shown on Fig. 3.3. It is located near the top edge of the scale, directly below the 1:20 scale. The numbering on this 1:200 scale is 0, 2 m, 4, 6, 8, 10, 12, 14 etc. Each large division represents 1 m or 1000 mm. Instead of reading the large divisions as 1 m, 2 m, 3 m, etc., count them off as 1000 mm, 2000 mm, 3000 mm, etc. Each of the small divisions represents 200 mm. Count them off on your scale as 200, 400, 600, 800, 1000, 1200, 1400, 1600, 1800, 2000. The 2000 dimension is marked on Fig. 3.3.

Check the 8000, 14500 and 21600 dimensions on Fig. 3.3. Then carefully check all the dimensions on drawing no. D9/1.

Using a 1:100 and Other Scales

Now look at the 1:100 scale. It is probably located near the edge directly underneath the 1:1 scale.

It should be divided into larger divisions numbered 0, 1 m, 2, 3, 4 and so on, which you should read as 1000 mm, 2000 mm, 3000 mm, etc. Each large division should be subdivided into ten small divisions representing 100 mm. This means, for example, that one large division and four small divisions represents 1400 mm.

A similar check can be made on all your other scales.

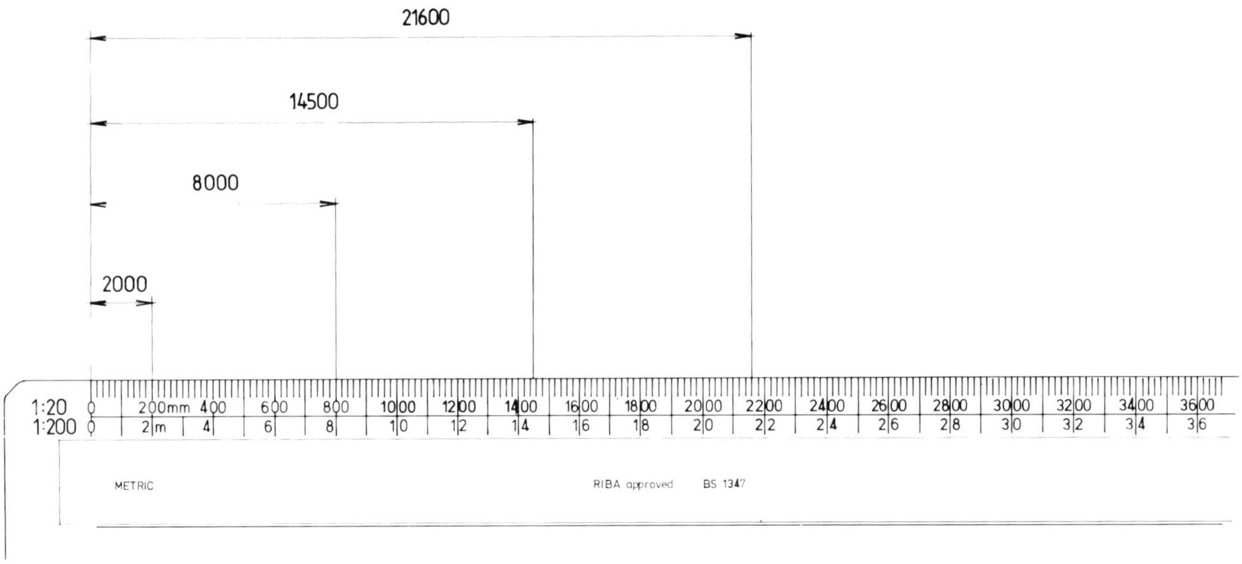

Fig. 3.3 Using a 1 : 200 scale

If your scale rule does not contain the scale you need, you can make use of a scale ten times smaller and obtain the correct dimension by dividing the figure on the scale by ten. For example use the 1:20 scale as a 1:2 scale (see Fig. 3.3). The dimension marked on this scale as 200 mm (i.e. two large divisions) becomes 20 mm and each small division becomes 2 mm (instead of 20 mm when the scale is 1:20).

Self Assessment Guide

In order to check whether you have understood this chapter you should now attempt to answer the following questions.

1 If you draw a building to a scale of 1:500, how much larger will the building actually be than than the size shown on your drawing?
2 What system of measurement does the British construction industry use?
3 How is the word 'metres' abbreviated?
4 How is the word 'millimetres' abbreviated?
5 How many metres are there in each of the following dimensions? 2000 millimetres, 7750 millimetres, 14500 millimetres.
6 How many millimetres are there in each of the following dimensions? 1.5 metres, 3.7 metres, 14.5 metres.

Now look at drawing no. D13/4 and check all the dimensions on this drawing with your 1:5 scale.

Now look at drawing no. D10/1 and check all the dimensions on this drawing with your 1:50 scale.

Now look at drawing no. D9/1 and check all the dimensions on this drawing with your 1:200 scale.

If you cannot answer any of the above questions, or relate the scales to the figured dimensions, carefully reread Chapter 3 until you are able to do so.

4

Lettering and Dimensioning

Introduction

One of the most important stages in producing a construction drawing is the lettering and dimensioning of the drawing. Every drawing needs a title, and often subtitles are required. In addition in order to make the drawing easier to understand and more useful to the builder and others, descriptive notes and dimensions will generally be required.

Quality

Faulty or poor quality lettering and dimensioning will at best spoil the appearance of the drawing, and at worst lead to mistakes being made at the construction site. It is important therefore that all work be done to a high standard and to achieve this practise in lettering and dimensioning will be necessary.

Methods of Lettering

Common methods of lettering include freehand lettering, stencilled lettering and dry transfers. Stencils and dry transfers are mentioned in Chapter 18, but in the early stages of their careers it is suggested that students use freehand lettering on their drawings.

Advantages and Disadvantages of Freehand Lettering

Freehand lettering is the cheapest way of annotating drawings and is generally the quickest method. It can be used for pencil or ink drawings and gives a distinctive appearance to a drawing compared to the mechanical look of drawings annotated with stencils or dry transfers.

Its disadvantage is that if more than one draughtsperson is engaged on a project the set of drawings will lack the uniformity achieved by the other two methods mentioned.

Requirements of Letters and Numerals

The first requirement of lettering is that it should be easy to read, both in the good environmental conditions of the drawing office, and the generally more difficult environment of the construction site. This is true of words, but even more important with figures, for if dimensions on a drawing are misread this can lead to mistakes being made during the construction process.

In order to assist legibility characters, whether letters or numerals, should be uniform and suitably formed. They should also be large enough, with spacings between individual letters making up words, and spaces between words, arranged to assist legibility. This means striking the right balance between making the letters and words too cramped, or too widely spaced.

Lettering should be neat and attractive to look at. This is more likely to be achieved if a simple style of letters and numerals are used.

It is also important to choose an appropriate size of letters and numerals. Consideration needs to be given to the fact that drawings are sometimes microfilmed, and reduced copies may be used on occasions, or drawings may be looked at on a microfilm viewer.

All lettering should be written parallel to the bottom edge of the drawing. Lettering should not be underlined.

Types of Letter

The two main groups of letters are 'CAPITAL LETTERS' and 'lower-case letters'. The use of lower-case letters is generally restricted to notes, but capital letters can be used for both notes and titles. The drawings in this book illustrate typical usage of the two types of letters. However, as it is easier to produce legible capital letters than lower-case letters, it is suggested that initially you use only capital letters on your drawings.

There are various styles of lettering, including Roman, Gill Sans and script lettering. A further subdivision is vertical lettering and inclined lettering. Where inclined lettering is used the slopes should be kept consistent at about 75 degrees.

Drawing no. D4/1 illustrates Gill Sans lettering, showing capital letters and numerals. An example of lower-case is also illustrated at the bottom of the page. This illustration provides some information on the proportions and other features of each letter and numeral. When you have finished studying this chapter it is suggested that you copy this information on a sheet of A4 paper (see Exercise 4.1).

Need for Guide Lines

It is important that all letters be formed between guide lines. A lower and upper guide line should be drawn as lightly as possible so that you can just see them. Drawing them takes a little extra effort but is worth the trouble.

Size of Letters

Most of the drawings which you will prepare as exercises in this book will be drawn on A4 or A3 paper. It is suggested that the capital letters and numerals have the following heights.

A4 sheets
Main titles 5 mm
Subtitles including room names 3.5 mm
Notes and dimensions 2.5 mm.

A3 sheets
Main titles 7 mm
Subtitles 5 mm
Room names 3.5 mm
Notes and dimensions 2.5 mm.

As a general rule the thickness of the lines forming the characters should be one tenth of the height of the letters. For example a 2.5 mm high letter should have lines 0.25 mm thick. Initially you will be governed by the pens and pencils available. It is suggested that the 2.5 and 3.5 mm high letters be written using 0.25 pens and fine-lead pencils, and that 5 and 7 mm high letters be written using 0.5 mm pens and pencils.

Spacing of Letters

The spacing of letters and words are governed by the need to achieve a pleasant appearance, and the need for legibility, including when drawings are microfilmed.

In practice this means that a minimum space of two pen thicknesses should be left between adjoining letters in a word. Thus 5 mm high letters, written with 0.5 mm pens, will have a minimum space of 1 mm

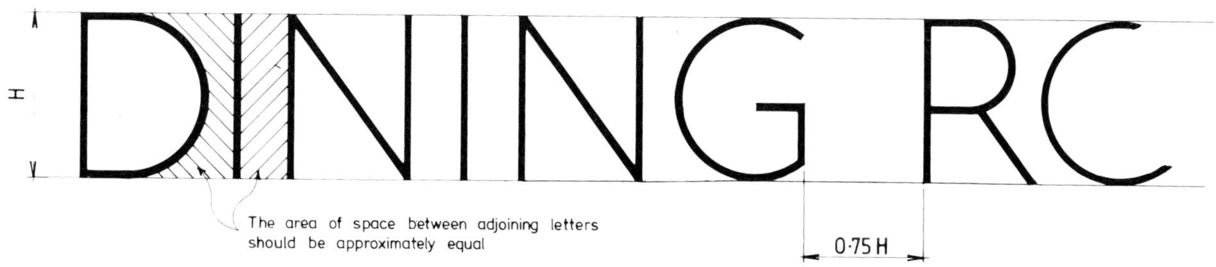

Fig. 4.1 Lettering

between adjoining letters in a word. The total area of space between adjoining letters in a word should be equal.

The clear space between adjoining words should be about three quarters of the height of the letters.

The space between any two lines of lettering should be a minimum of half the height of the letters.

The above principles are illustrated in Fig. 4.1.

Need to Practise Lettering

The information given so far in this chapter is intended to help the novice, and should not be adopted in an arbitrary manner. With practice beginners will develop their own personal style of lettering. They will also acquire the ability to space the letters and words so that the written information is both legible and attractive. This ability will only come with practice, and students are strongly advised to conscientiously undertake the exercises at the end of this chapter. This will be time well spent, as good lettering is an essential ingredient of a drawing of a professional standard.

Annotation

The appearance of drawings is improved if the descriptive notes are arranged as regular shaped panels. Each specific item should form an individual paragraph.

Dimensioning

It is important that all drawings are fully dimensioned, so that the builder and others know the required size of every part of the building. It is sensible however not to duplicate dimensions, as this makes the drawing unnecessarily crowded.

Dimension lines should be unbroken lines. They can be terminated at their ends by open arrowheads, solid arrowheads, oblique strokes, dots or circles. All of these methods are illustrated in drawing no D4/2. Open arrowheads are widely used in architectural drawings, particularly for basic dimensions. Angles of open arrowheads should be 45 degrees or less. Solid arrowheads are also popular, particularly in engineering drawings. This type of arrowhead has angles of 10 to 45 degrees, and the length of each arrowhead is typically 3 mm. Oblique strokes are sometimes used for runs of small dimensions. Dots are useful when the dimensions are too small for arrowheads to be used. Circles fulfil a similar purpose, but they are less common. Any of the terminations shown may be used, but it is best not to mix the various types on the same drawing. It is suggested that the beginner uses the open arrowhead initially.

Thin lines called projection lines, or projectors, should extend from about 2 mm away from the part of the object being dimensioned to just beyond the dimension line termination.

Dimensions should be written in numerals about 2.5 mm high, located in the centre of the dimension line, parallel to the line, and close to it. As previously mentioned there is no need to add the mm suffix after the number. In the case of vertical dimensions they should all be the same way round so that they can be read from the right hand side of the drawing.

Horizontal Dimensions

Horizontal dimensions should, where possible, be indicated on plans rather than on elevations. Where feasible dimension lines should be located outside the building or object rather than inside it.

Vertical Dimensions

Vertical dimensions should, where possible, be indicated on sections rather than elevations. All vertical dimensions of a building should relate to a site datum. The site datum is a fixed vertical level on the site, and for convenience is often set at the ground floor level of the building under construction.

Different members of the building team tend to follow different practices regarding the measuring points for vertical dimensions. The client is concerned with clear storey heights—i.e. the dimension between the finished floor level and the finished ceiling levels. Architects will invariably give the finished floor level (FFL) on their drawings. Site staff work initially to the structural floor level (SFL). Carpenters working on a building having a timber roof will be interested in the level of the underside of the ceiling joists, rather than the level of the ceiling finish. Structural engineers need to know the structural floor level (SFL), and their vertical dimensions will generally be measured from SFL to SFL.

Drawing no. D11/1 shows a cross section through a small bungalow. Key levels are the underside of the foundations, the top of the structural floor slab (SFL), and the underside of the ceiling joists, but the FFL is also given.

Final Point

The final point about dimensions is an obvious one. Mistakes in dimensioning can be expensive both in terms of time and money, so it is worth checking and double checking to ensure that there are no errors. If any dimension is not drawn to scale, the letters 'NTS' (not to scale) should be written after the dimension.

Exercise 4.1 Copying Gill Sans lettering

Requirement
Make an exact copy of drawing no. D4/1, in pencil, on a sheet of white A4 paper.

Preliminaries
Begin by drawing a 10 mm border around the edge of the A4 paper. Add the title panel at the bottom of the drawing. This is in the form of a 20 mm wide continuous horizontal strip. Subdivide the title panel to provide spaces for the title, name, date, scale and drawing number.

Lettering
Draw all the guide lines. You can do this by measuring their positions on drawing no. D4/1 in the book, and transferring these measurements on to your own A4 sheet.

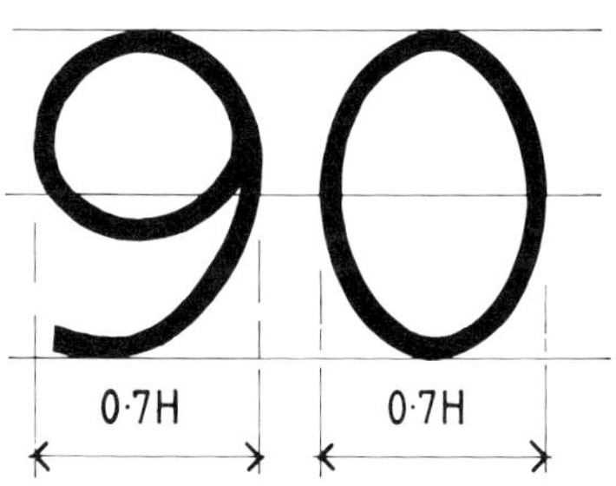

aabcdefghijklmnopqrstuvwxyz [ ] ( ) ! ?

ABCDEFGHIJKLMNOPQRSTUVWXYZ 01

lettering guides are produced as a 'sans serif' script

GILL SANS LETTERING	NAME DATE SCALE N.T.S.	D4/1

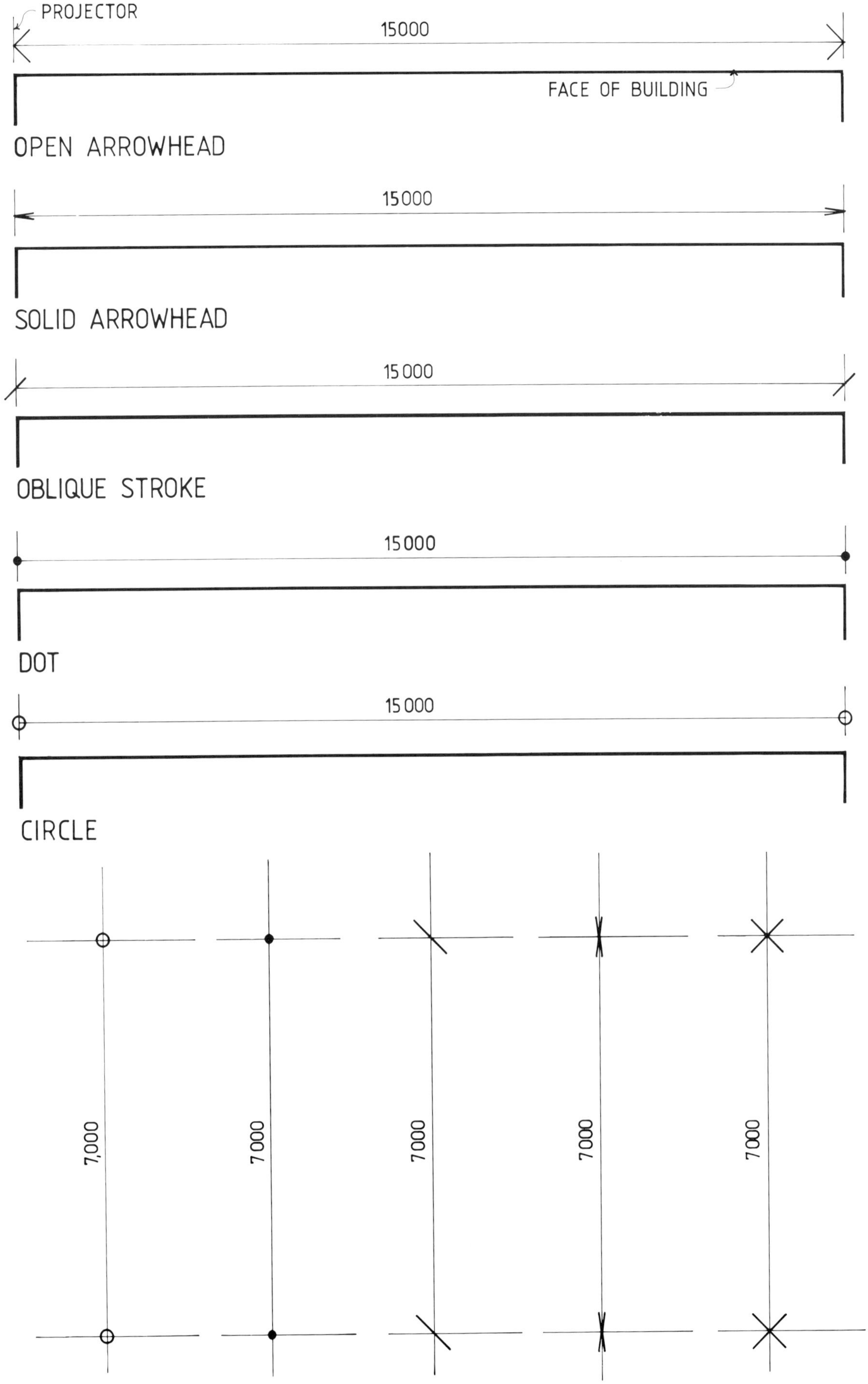

DIMENSION LINES

NAME

DATE

SCALE

D4/2

Lightly draw in the shapes of all the letters and numerals, using your straight edge/tee-square, set-square and circle template.

Add all dimensions and notes to your drawing.

When you are satisfied that you have accurately copied drawing no. D4/1, firm in all the lines.

Exercise 4.2 Practising lettering

Requirement
Practise Gill Sans lettering 5 mm high, in pencil, on a sheet of white A4 paper.

Preliminaries
Draw the border and title panel in the same way described for Exercise 4.1.

Lettering
Draw horizontal guide lines 5 mm apart over the whole of the A4 sheet. Carefully form freehand characters—both capital letters A to Z and numerals 0 to 9—in every other pair of guide lines.

Exercise 4.3 Copying dimension lines

Requirement
Make an exact copy of drawing no. D4/2, in pencil, on a sheet of white A4 paper.

Preliminaries
Draw the border and title panel in the same way described for Exercise 4.1.

Dimensioning
Draw all the faint construction lines first, and then draw over them with firm final lines. Draw the dimension lines first, working from top to bottom, and from left to right. Then add the lettering and figures. The positions of the dimension lines can be measured from drawing no. D4/2 in the book.

Exercise 4.4 Practising pencil tracing

Requirement
Trace your drawing produced for Exercise 4.1, in pencil, on a sheet of A4 tracing paper.

Exercise 4.5 Practising ink tracing

Requirement
Trace your drawing produced for Exercise 4.3, in ink, on a sheet of A4 tracing paper.

Self Assessment Guide

In order to check whether you have understood this chapter you should attempt to answer the following questions.

1 What are the dangers of faulty or poor quality lettering?
2 What are the advantages of freehand lettering?
3 What is the first requirement of lettering?
4 State what height of lettering you will use while working through the exercises in this book, for each of the following. Notes; dimensions; room names; subtitles; main titles.
5 What governs the spacing of letters and words?
6 What size pen is suitable for 2.5 mm high lettering?
7 State a type of dimension termination which is widely used for architectural drawings.
8 What are projection lines or projectors in relation to dimensioning?
9 Are vertical dimensions generally indicated on elevations or sections?

When you have attempted all the above questions, reread the appropriate part of the chapter and check you know the correct answer to every question.

You should now check that you have gained the necessary skill in lettering and dimensioning. You should be able to answer YES to each of the following questions.

Lettering

1 Have you used guide lines for all your lettering?
2 Are the guide lines very faint—i.e. barely visible?
3 Do all the letters exactly fit between the guide lines?
4 Are all the letters easy to read?
5 Do you consider the letters are attractive to look at?
6 Are the shapes and proportions of the letters similar to those shown on drawing no. D4/1?
7 With regard to the spacing of the letters and words have you followed the principles illustrated on Fig. 4.1?
8 Are similar letters identical? E.g. are all the 'A's exactly the same?

Dimensioning

9 Are all your dimension lines smooth, even and of the same thickness throughout their length?
10 Are all the dimensions legible, neat and the same size?
11 Do all the projection lines extend from about 2 mm away from the object being dimensioned to just beyond the dimension line termination?
12 Are all the dimensions written in the centre of the dimension line, parallel to the line and close to it?
13 Are all the vertical dimensions arranged the same way round so that they can be read from the right hand side of the drawing?

5

Graphic Conventions

Introduction

Construction drawings are a means of communication between the various members of the building team, and it is important that they employ a common graphical language. It helps to achieve this if agreed standards are followed in respect of lines, hatchings and symbols, etc. A good convention should be simple, distinctive and easy to draw.

Lines

Lines vary in thickness and form according to their purpose and importance.

Construction lines have already been mentioned. They are setting-out or guide lines, and they should be drawn as light as possible. They are generally covered over by the stronger final lines.

Thick active lines are continuous, and used to indicate important parts of structures such as the outside and inside faces of walls; the faces of reinforced concrete members and ground levels.

Thin active lines are continuous, and used to show items drawn as plan views (as opposed to sectional plans) and as elevations; also to define less important items shown in section.

Hidden lines are broken lines, and can be either thick or thin depending on their importance. They show work which is not visible—e.g. the position of beams on a floor plan.

Centre lines are thin chain dotted lines and are used as the name implies, to show the centre of things—e.g. the centre of a beam.

Break lines are thin continuous lines with a zig zag in them to show a break in the continuity of the line or view.

Drain and other underground pipe lines may be shown by a thick chain line or a thick continuous or broken line. In the case of underground drains, arrowheads are often added to show the direction of flow.

Dimension lines and projectors were dealt with in Chapter 4. Dimension lines are illustrated in drawing no. D4/2, and the remaining types of line in drawing no. D5/1.

Section Lines or Planes

A section is a view of a building or object obtained by making an imaginary cut through it. This is explained further in Chapters 6 and 11, but is mentioned now because of the need for you to be able to recognise what a section line looks like. Sometimes called a section plane, it shows the position where the imaginary cut is made. The line itself is a chain dotted line, with the line terminated by arrows which point in the direction of the viewing. The arrows are labelled with capital letters—the letters I and O are not used for this purpose. A section line or plane is shown on drawing no. D5/1, and also on drawing no. D10/1 (see pages 32 and 75).

THICK ACTIVE LINES
defining main outlines of structures in section

THIN ACTIVE LINES
defining outlines in plan and elevation

HIDDEN LINES – THICK OR THIN
showing work not visible or work to be removed

CENTRE LINES – THIN LINES

BREAK LINES – THIN LINES
for break in continuity

UNDERGROUND (INFRASTRUCTURE) – THICK LINES
drainage runs, pipe lines, cables etc.

SECTION PLANES – THICK AND THIN
thick lines at ends and changes of direction only, thin elsewhere

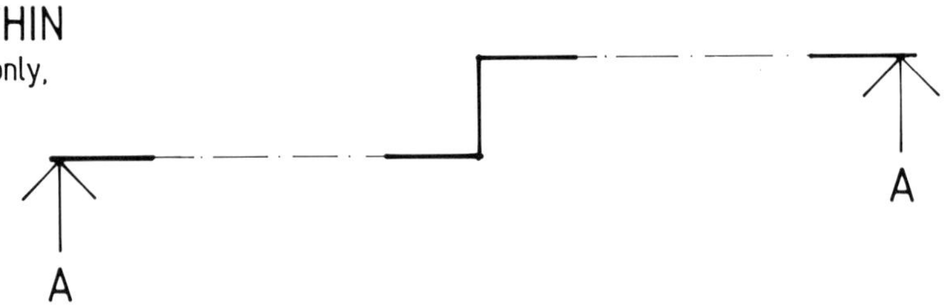

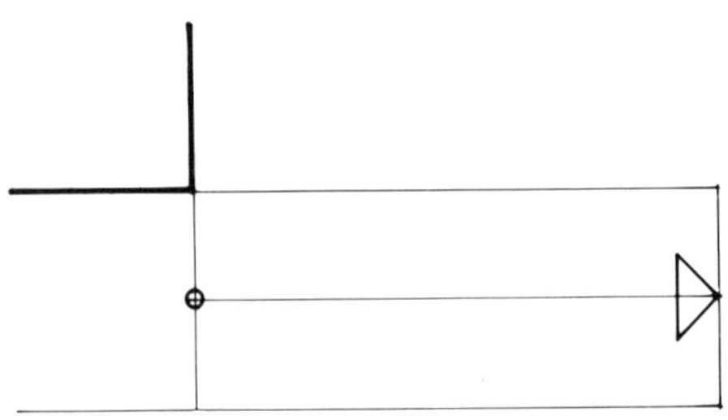

RAMP arrow indicates direction of fall from ⌖

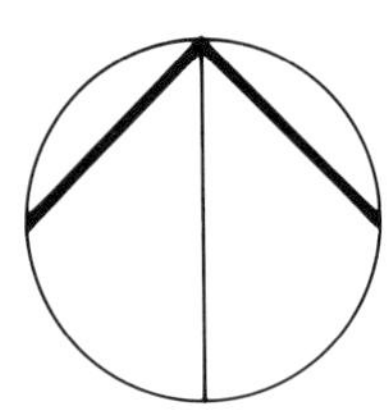

NORTH POINT

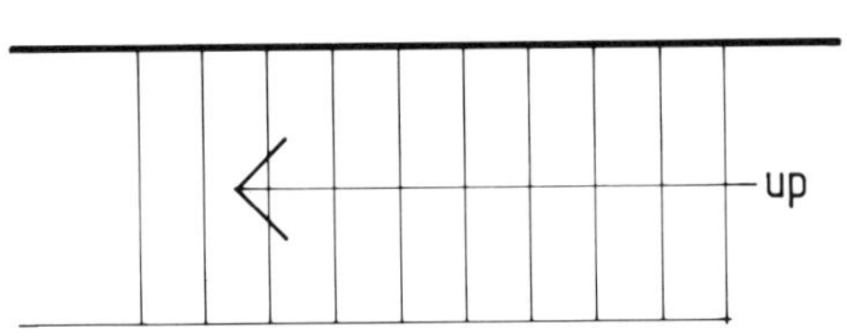

STAIRS arrow indicates direction of travel

LINES, STAIRS and NORTH POINT	NAME DATE SCALE	D5/1

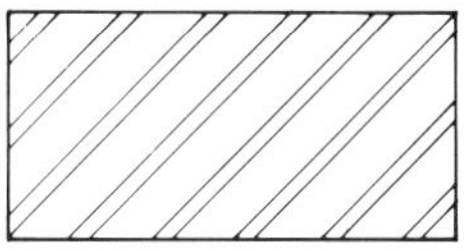

BRICKWORK

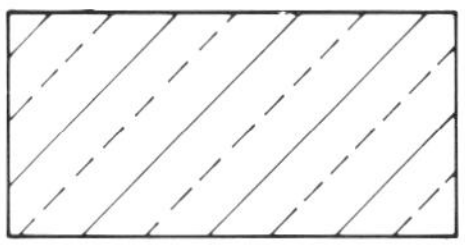

STONEWORK

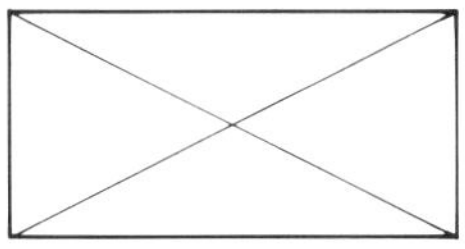

TIMBER – SAWN

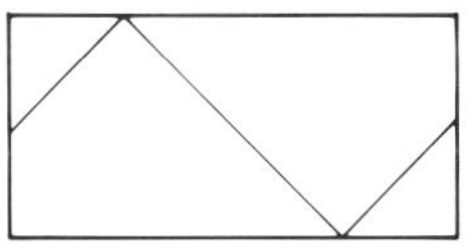

HARDCORE

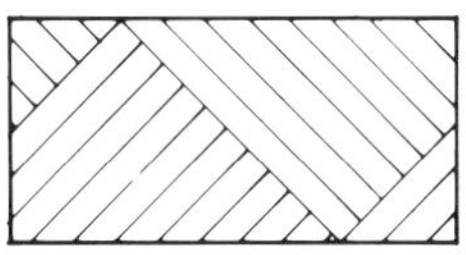

SUBSOIL

SCREED / RENDER / PLASTER

PLYWOOD

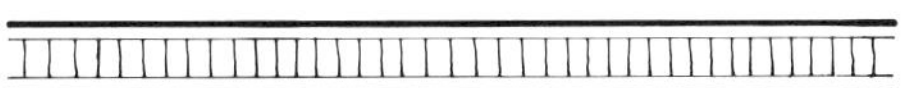

BLOCKBOARD

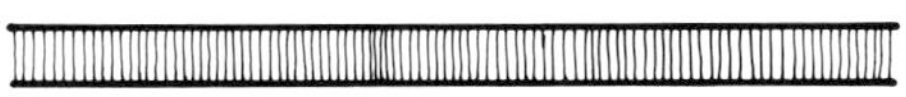

ASPHALT

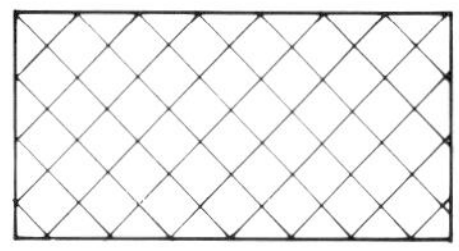

BLOCKWORK

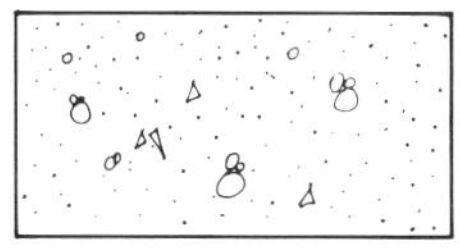

CONCRETE

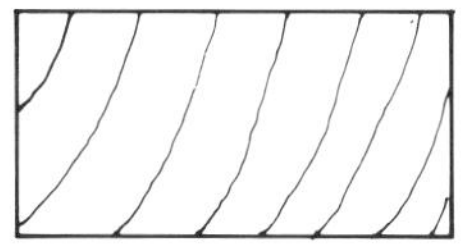

TIMBER – PLANED

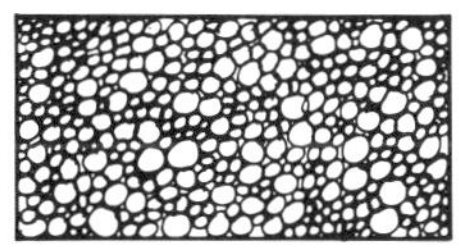

GRANULAR FILL

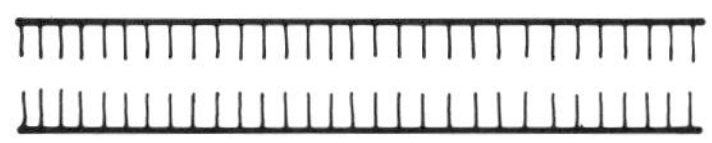

TOPSOIL

STEEL SECTION

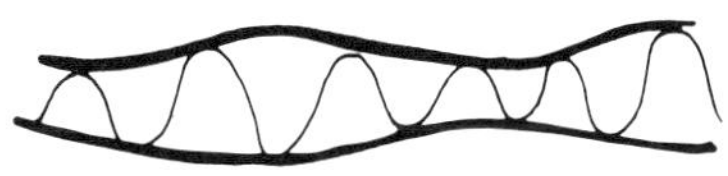

INSULATION QUILT

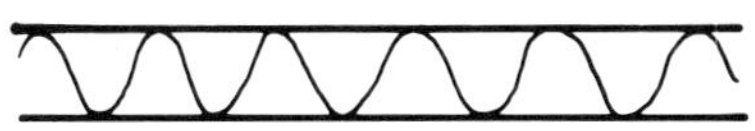

INSULATION BOARD

MATERIALS IN SECTION	NAME DATE SCALE	D5/2

Alternative methods for single leaves opening 90°

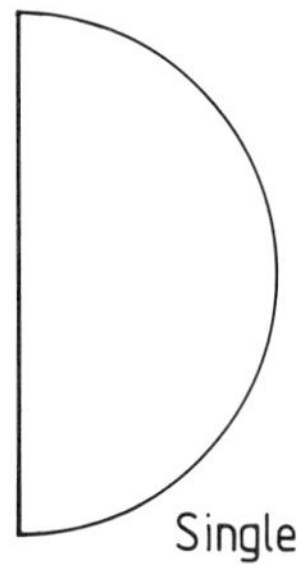

Single leaf opening 180°

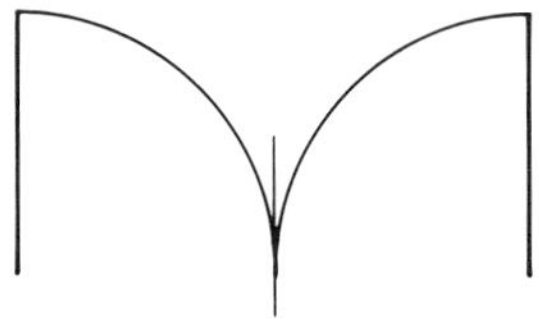

2 leaves each opening 90°

HINGED DOORS — PLANS

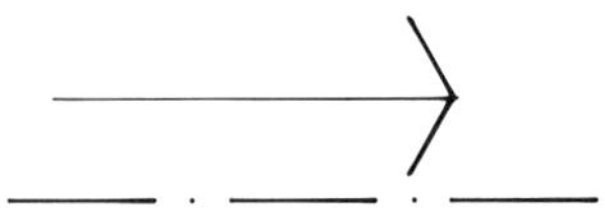

Straight sliding

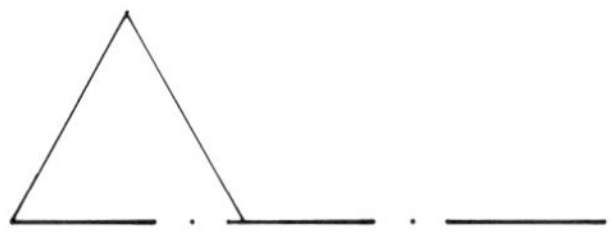

Leaves end hung

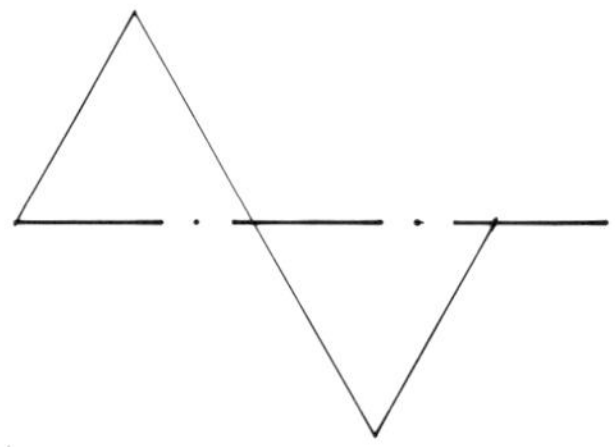

Leaves centre hung

SLIDING DOORS — PLANS

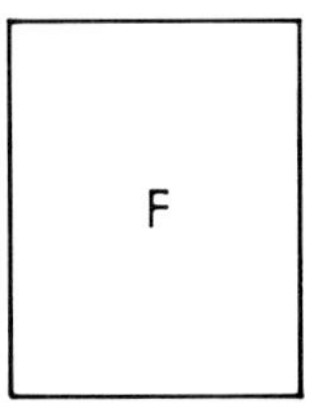

FIXED

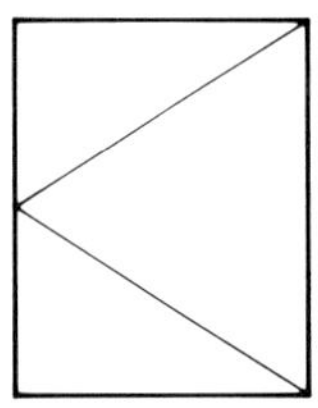

SIDE-HUNG

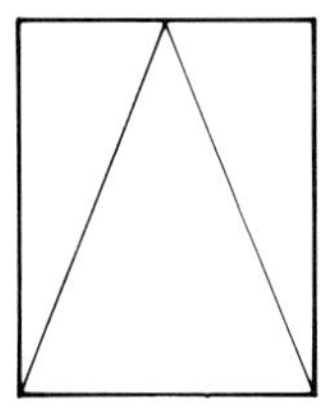

TOP-HUNG

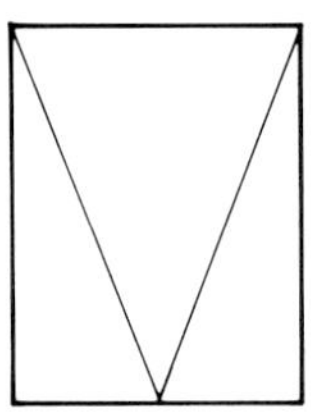

BOTTOM-HUNG

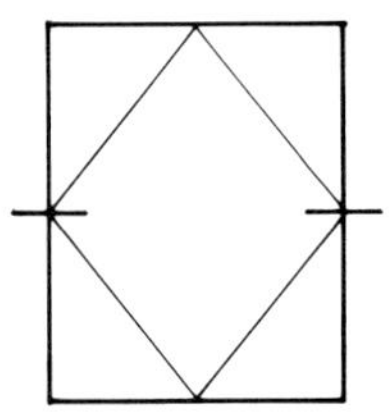

PIVOT – HORIZONTAL

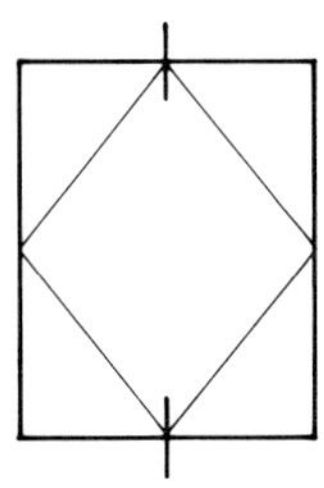

PIVOT – VERTICAL

WINDOWS — ELEVATIONS

DOOR / WINDOW SYMBOLS

NAME

DATE

SCALE

D5/3

Staircases and Ramps

It is usual to show the 'up' direction of a flight of stairs, or a ramp, drawn on a plan by means of an open arrowhead. A closed arrowhead may be used to indicate the 'down' direction of a ramp, or slope. Both of these cases are illustrated in drawing no. D5/1.

Representation of Materials

In sectional views of a building, the parts of the structure which are cut by the section plane may be hatched to indicate the nature of the materials used. Common examples of hatching for construction materials are given in drawing no. D5/2. The decision whether or not to use cross hatching is influenced by the scale of the drawing.

Doors

There are standard ways of indicating on plans the opening methods for doors—i.e. whether they are swing doors or sliding doors—and the direction in which they open. These are illustrated in drawing no. D5/3.

Windows

The opening methods for windows are generally indicated on the elevations. They are illustrated in drawing no. D5/3.

Symbols

There are a wide range of standard graphic symbols available to indicate the position of various components, and related information. Some common examples are given below, but reference needs to be made to BS 1192 for full details of the recommendations for symbols and other graphic conventions.

Drawing no. D5/4 gives details of items commonly used on architectural drawings.

Drawing no. D5/5 gives details of some of the many symbols used to indicate building services items.

Orientation of Plans

North points are generally shown on key plans, site plans, block plans and sometimes floor plans, to indicate the position of north relative to the site or building. The point of the arrow should be drawn to face north.

Drawing no. D5/1 shows an example of a north point.

Abbreviations

Words are sometimes abbreviated on drawings. A list of common examples is given below.

AB	Air brick
AC	Alternating current
AC	Access cover
AG	Access gully
BIG	Back inlet gully
Blw	Blockwork
BM	Bench mark
Brk	Brickwork
BS	British Standard
C/C	Centre to centre
CE	Cleaning eye
CI	Cast iron
CL	Cover level
CONC	Concrete
CWS	Cold water service
D	Diameter

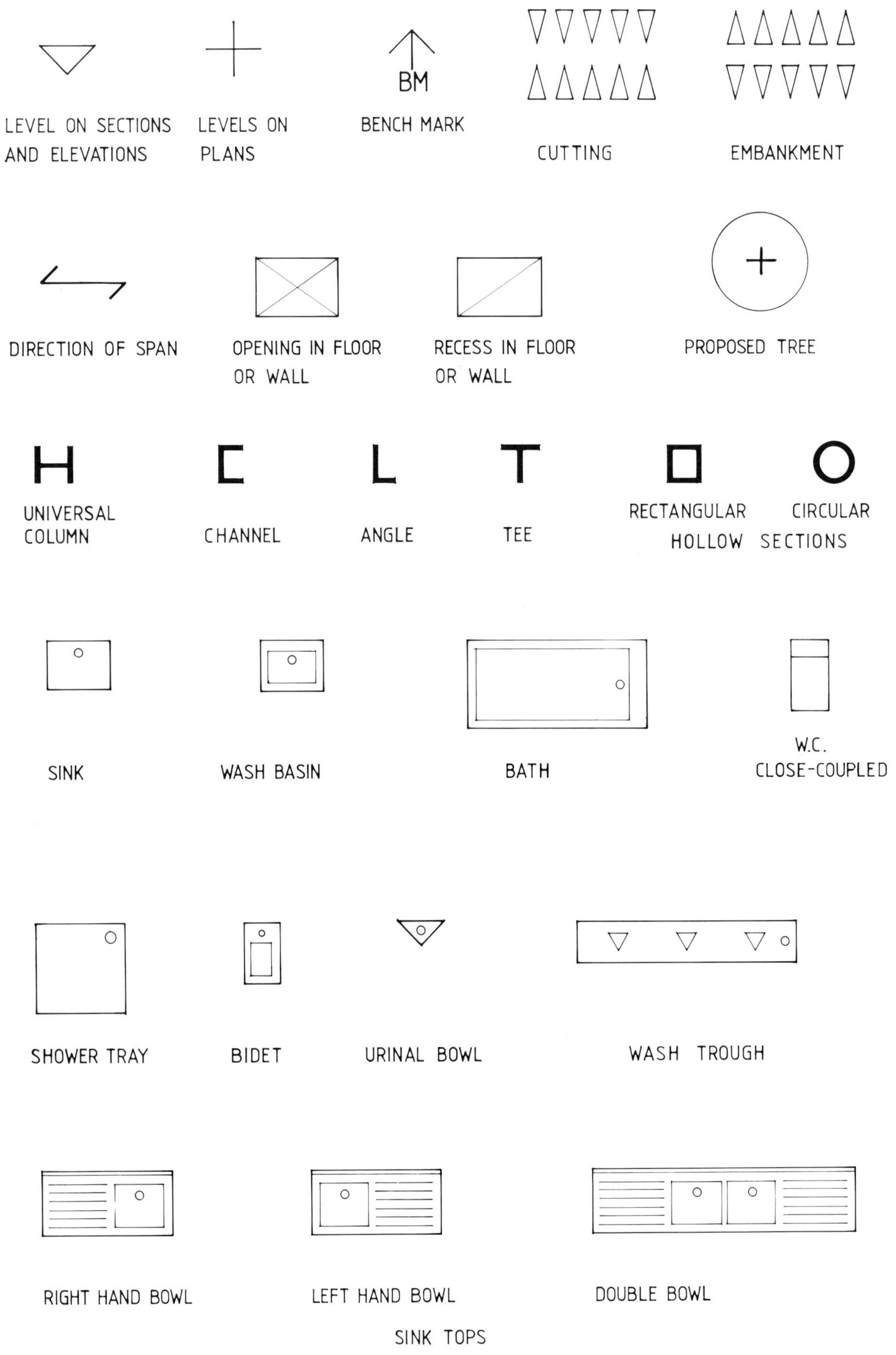

ARCHITECTURAL SYMBOLS	NAME	D5/4
	DATE	
	SCALE	

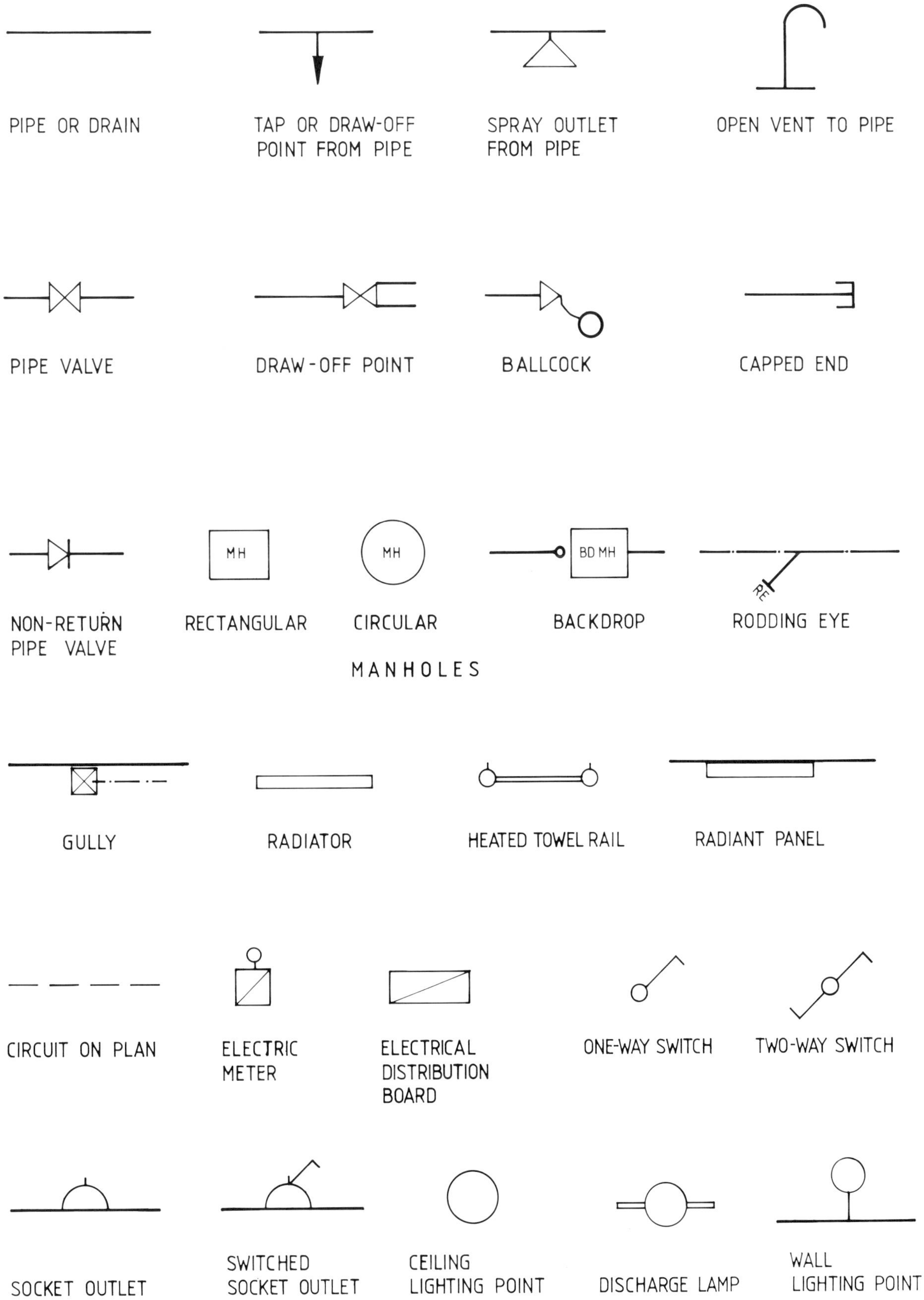

SERVICES SYMBOLS	NAME DATE SCALE	D5/5

DC	Direct current
DPC	Damp proof course
DPM	Damp proof membrane
F	Fixed (window)
FFL	Finished floor level
FG	Floor gully
FIG	Figure
G	Gully
GL	Ground level
HWS	Hot water service
IC	Inspection cover
IL	Invert level
LB	Lavatory basin
LH	Left hand
MH	Manhole
MIN	Minimum
MS	Mild steel
NO	Number
NTS	Not to scale
RC	Reinforced concrete
RH	Right hand
RWP	Rain water pipe
S&VP	Soil and vent pipe
SP	Soil pipe
SW	Surface water
T&G	Tongued and grooved
TBM	Temporary bench mark
UPVC	Unplasticised polvinyl chloride
VP	Vent pipe
WB	Wash basin
WC	Water closet

Exercise 5.1 Lines, stairs and north point

Requirement
Make an exact copy of drawing no. D5/1, in pencil, on a sheet of white A4 paper.

Procedure
Draw a 10 mm border around the edge of the sheet. Add a title panel at the bottom of the drawing, subdivided to provide spaces for the title, name, date, scale and drawing number.

Draw all the construction lines first. Then firm in the final lines. Finally add the lettering.

Exercise 5.2 Materials in section

Requirement
Make an exact copy of drawing no. D5/2, in pencil, on a sheet of white A4 paper.

Procedure
Follow the same procedure described for Exercise 5.1.

Exercise 5.3 **Door and window symbols**

Requirement
Make an exact copy of drawing no. D5/3, in pencil, on a sheet of white A4 paper.

Procedure
Follow the same procedure described for Exercise 5.1.

Exercise 5.4 **Symbols**

Requirement
Redraw all the symbols contained on drawing numbers D5/4 and D5/5, in pencil, on a sheet of white A3 paper.

Procedure
Draw a 10 mm border around the edge of the A3 sheet. Add a title panel at the bottom of the drawing. This is to be in the form of a 20 mm wide continuous horizontal strip. Subdivide the title panel to provide spaces for the title, name, date, scale and drawing number, as indicated on Fig. 2.2.

Draw the construction lines for all the symbols. Draw the guide lines for all the lettering, including the title. Firm in the final lines. Write all the lettering, including the title.

Exercise 5.5 **Practising ink tracing**

Requirement
Trace your drawing produced for Exercise 5.4, in ink on a sheet of A3 tracing paper.

Self Assessment Guide

In order to check whether you have understood this chapter you should now attempt to answer the following questions.

1 What is the reason for following agreed standards in respect of types of lines, hatchings, symbols etc.?
2 What are construction lines?
3 What are thick active lines used for?
4 What are thin active lines used for?
5 What are hidden lines?
6 What are break lines?
7 What is a section plane, and how is it indicated?

When you have attempted all the above questions, reread the appropriate part of the chapter and check you know the correct answer to every question.

You should now check that you have completed the exercises to the correct standard.

Check that you have correctly copied the seven different types of lines shown on drawing no. D5/1, and that the lines described as 'thick' are about 0.5 mm, and the lines described as 'thin' are about 0.25 mm.

Check that the information you have copied in respect of stairs, ramps and the north point is exactly the same as shown on drawing no. D5/1.

Check that you have correctly copied from drawing no. D5/2 on to your drawing sheet, the correct conventional representation for all the materials in section.

Check that the information you have copied in respect of doors and windows is exactly the same as shown on drawing no. D5/3.

Check that the information you have copied in respect of symbols is exactly the same as shown on drawing numbers D5/4 and D5/5.

Check that all the lines you have drawn are smooth, even and of constant thickness throughout their length.

Check that all the slanting lines are drawn at the correct angle—generally at 45 degrees.

Check that at junctions between horizontal and vertical lines they slightly cross each other, or exactly meet.

Check that you have used guide lines for all lettering and that your letters are legible, neat and of constant height.

Check that the borders and title panels have been correctly drawn on all your sheets.

6

Orthographic Projection

Introduction

Buildings, and the materials and components of which buildings are constructed, are three dimensional. That is to say they have length, width and height. It is possible to draw a picture of a building or object to show these three dimensions, and two such methods are described in Chapter 7. Generally however in construction drawing the method used to describe buildings or objects pictorially is called *orthographic projection*. This method uses views termed plans, elevations and sections, which have only two dimensions.

Plans

A plan is generally a view of a building or object obtained by looking downwards on it. Drawing no. D6/1 illustrates a concrete building block used in the construction of walls. A three dimensional picture of the building block is shown at the bottom right hand corner of the sheet. Near the bottom left hand corner is a plan of the building block. This shows the view you would get if you looked down on the block. The outer edges of the block are indicated by two horizontal lines and two vertical lines, forming a rectangle. There are three smaller rectangles inside this larger one, showing that the block is hollow and not a solid concrete block.

Drawing no. D6/2 illustrates a stone coping which is used to finish the top of a pier. Near the bottom left hand corner of the sheet is a plan of the coping—i.e. the view you would get if you looked down on to it. Again the four outer edges of the coping are indicated by two horizontal and two vertical lines. The coping is sloping, or weathered, at the top so as to throw off the rain, and this is shown by a central horizontal line and four lines slanting, in this case, at 45 degrees.

Drawing no. D6/3 illustrates a small garden store. In the bottom left hand corner of the sheet is a roof plan. This is the view you would get if you were to fly over the store. Because it is a very simple building the roof plan just consists of a rectangle. Inside this rectangle is a smaller rectangle made up of two horizontal and two vertical broken (to denote hidden) lines. These hidden lines indicate the position of the outer wall faces directly below the roof slab.

There is a floor plan to the right of the roof plan. Although it is generally referred to as a plan or floor plan, it is really a sectional plan. Imagine that a huge knife or saw cuts through the store half way up its height. If you move the top part of the store away, and look down at the remaining bottom part, you will see the floor plan shown on drawing no. D6/3. On such a plan you will not only see the external shape of the building, but also the thickness of the walls and the door.

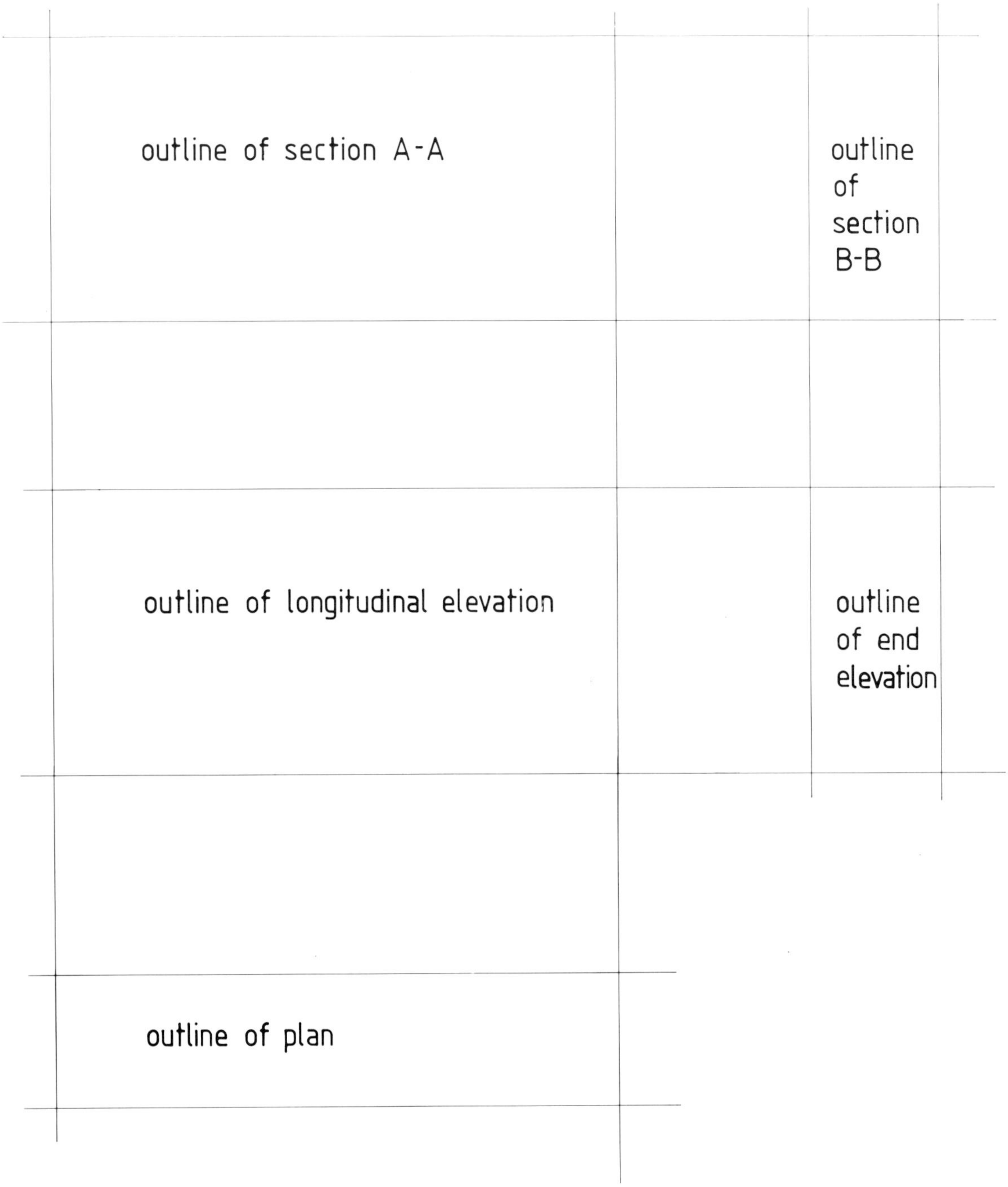

ORTHOGRAPHIC PROJECTION OF BUILDING BLOCK – FIRST STAGE	NAME DATE SCALE 1:5	D6/1 1st STAGE

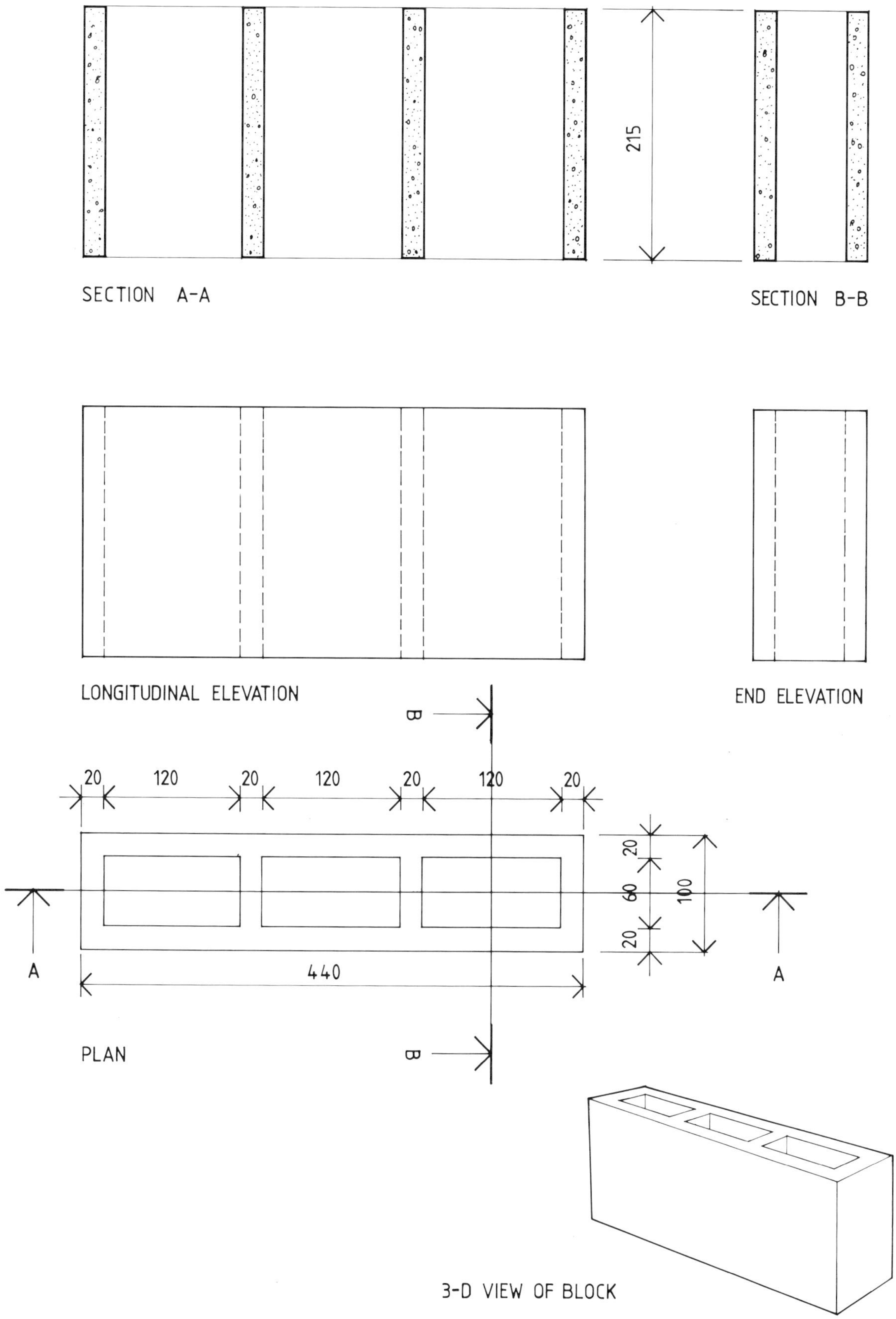

ORTHOGRAPHIC PROJECTION OF BUILDING BLOCK	NAME	D6/1
	DATE	
	SCALE 1:5	

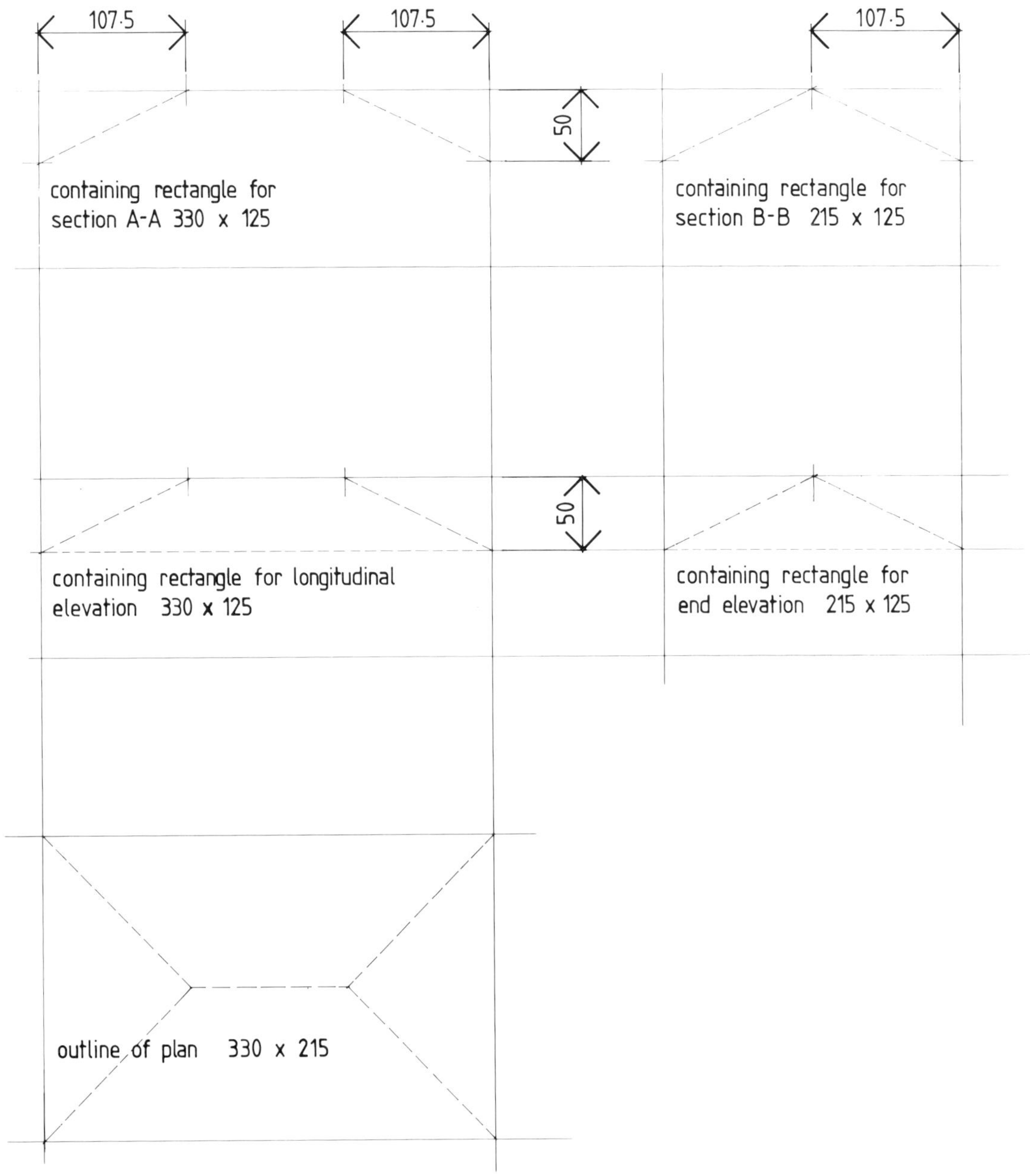

CONSTRUCTION LINES

Denotes 1st STAGE ————————

Denotes 2nd STAGE - - - - - - - - -

ORTHOGRAPHIC PROJECTION – COPING 1st and 2nd STAGE	NAME DATE SCALE 1:5	D6/2 1st + 2nd Stage

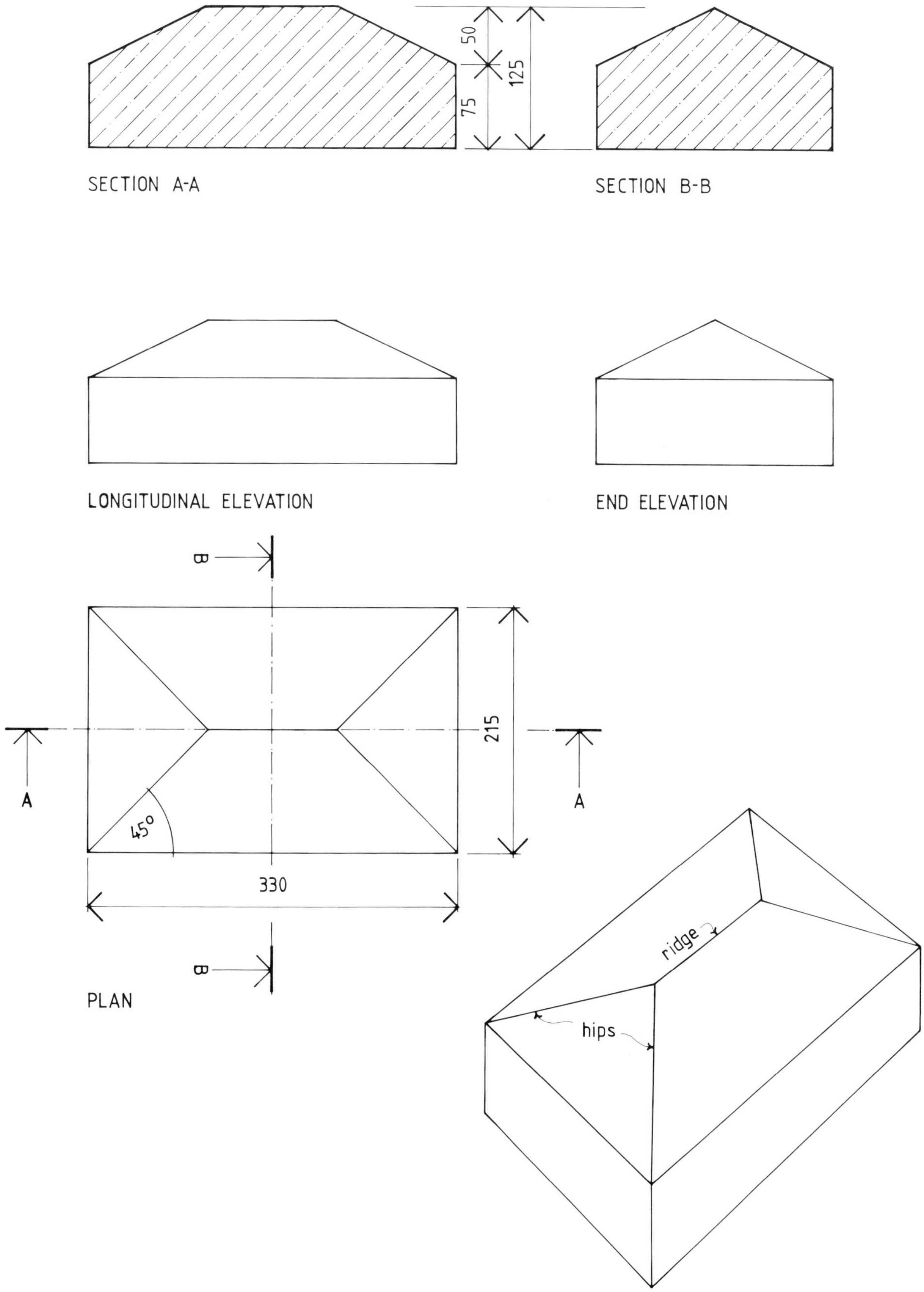

STONE COPING

NAME	
DATE	
SCALE 1:5	

D6/2

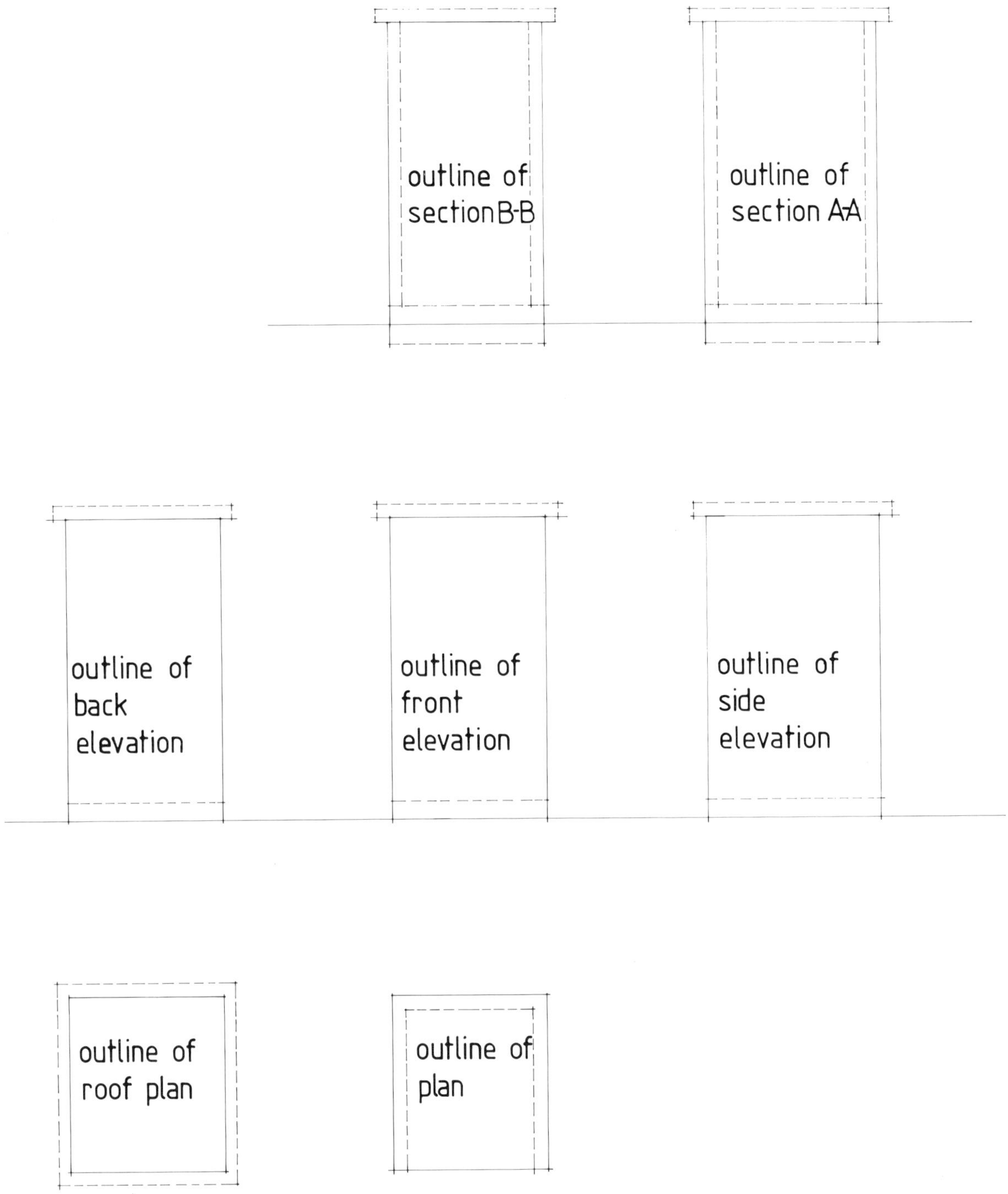

CONSTRUCTION LINES

Denotes 1st STAGE ————————

Denotes 2nd STAGE – – – – – – –

ORTHOGRAPHIC PROJECTION – GARDEN STORE 1st and 2nd STAGE	NAME DATE SCALE 1:50	D6/3 1st + 2nd Stage

SPECIFICATION

FLOOR	150 concrete slab on 100 compacted hardcore
WALLS	102·5 Rustic Fletton brickwork
ROOF	100 R.C. slab with proprietary bitumen-based finish
DOOR	915 × 2135 flush timber in 100 × 50 softwood frame

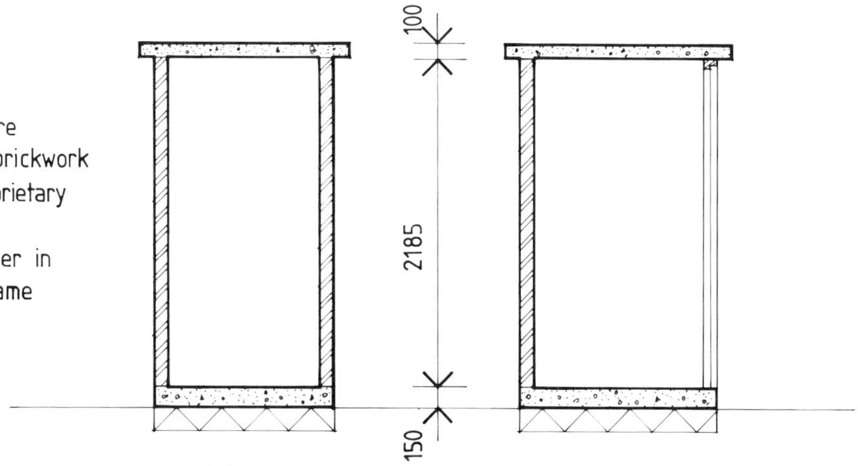

SECTION B-B   SECTION A-A

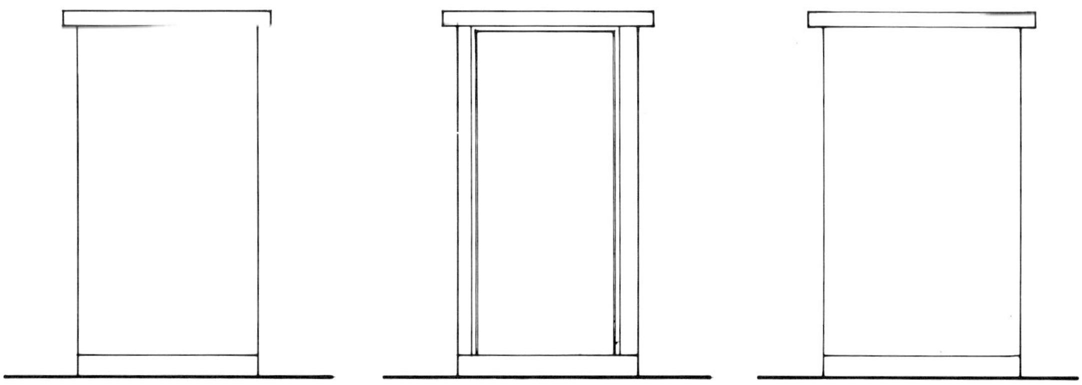

BACK ELEVATION   FRONT ELEVATION   SIDE ELEVATION

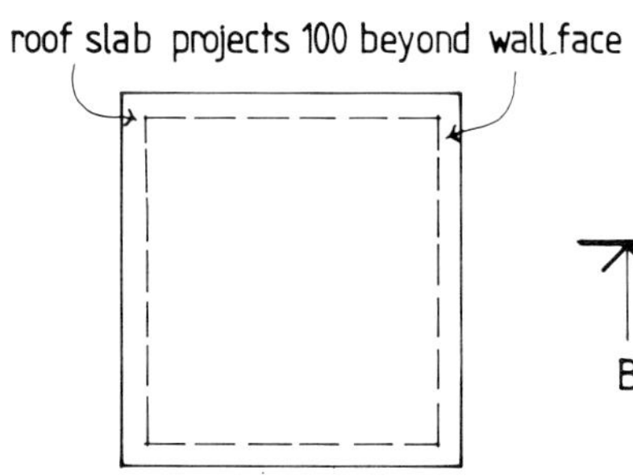

ROOF PLAN

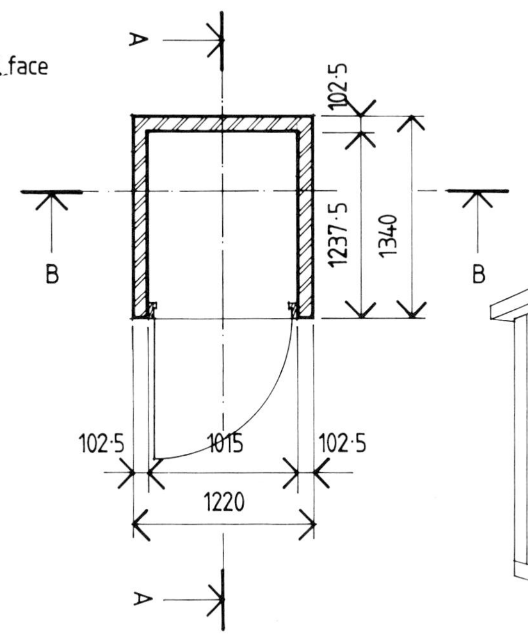

PLAN

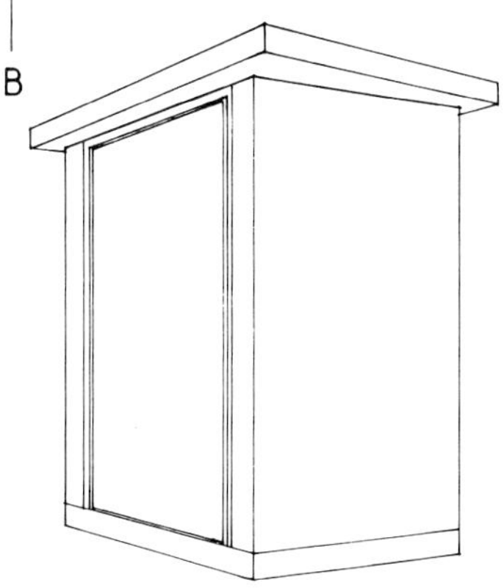

GARDEN STORE	NAME DATE SCALE 1:50	D6/3

Elevations

An elevation is a view you get if you look in a horizontal direction at the vertical sides, or faces, of a building or object. If you look at the two long sides of the concrete building block illustrated in drawing no. D6/1, the view you will get is of a rectangle which is fairly long—i.e. the length is much greater than the height. This would normally be called a longitudinal elevation. Broken (hidden) vertical lines on this elevation indicate the position of the holes in the block. If you look at the other two sides of the building block, again the view you have is of a rectangle, but this time its height is greater than its width. This would normally be called an end elevation.

The longitudinal elevation of the coping shown on drawing no. D6/2 consists of two parts. Firstly the vertical face which form a rectangle. Above this there is a regular trapezium having a shape similar to the longitudinal elevation of a hipped roof. The end elevation also has a rectangle to the lower part, with the upper part formed by a triangle. The reason for this is obvious from a study of the three dimensional illustration in the bottom right hand corner of drawing no. D6/2.

Similarly a study of the three dimensional illustration of the garden store on drawing no. D6/3 will show the reason for the arrangement of the front, back and side elevations.

Sometimes when drawing elevations of buildings, instead of using the terms *front* elevation, *back* elevation and *side* elevation, the terms *north* elevation, *south* elevation, *west* elevation and *east* elevation are used instead. This is useful when the two side elevations are different, or in cases such as on a large site, when it is not clear which is the front and which is the back of the building.

Sections

A section is a view of a building or object obtained by making an imaginary cut through it. In the case of drawing no. D6/1 if you were to cut through the concrete building block vertically with a saw or huge knife at the point A—A on the plan, and look at one of the cut faces, the view you would see would be as shown on section A—A. Similarly if you were to make the cut at the point B—B on the plan, the sectional view you would get would be as shown on section B—B.

In the case of the coping illustrated on drawing no. D6/2 if you were to make the cut at point A—A on the plan you would have a view as shown on section A—A. Similarly if you were to make the cut at the point B—B on the plan, the sectional view would be as shown on section B—B.

In the case of the garden store illustrated on drawing no. D6/3 if you were to make the cut at point A—A on the plan the sectional view would be as shown on section A—A. This shows the wall at the left hand side, then the space inside the store, and then the door on the right hand side. At the bottom of the section, the imaginary knife or saw cuts through the reinforced concrete floor slab laid on a hardcore bed, and at the top of the section the cut is made through the reinforced concrete roof slab. Section B—B is similar to section A—A except that there is a wall on both sides of the section.

Layout of Orthographic Projection Drawings

There are no hard and fast rules regarding the layout of the various views—plans, elevations and sections—used in orthographic projection. In the case of a small building all the views will generally be arranged on a single sheet. It is usual to place the plans at the bottom of the drawing sheet, with the front elevation directly above the ground floor plan. It is a good general rule to arrange the sections horizontally in line with the elevations. As mentioned under 'General Drawing Strategy' in Chapter

2, the drawing should be arranged with spaces between the different views so it does not look too crowded.

When preparing the drawings for larger buildings, plans will be arranged on one or more sheets; often each floor plan will be drawn on a separate sheet. The elevations will also be drawn on one or more sheets, as will the sections.

Exercises Generally

In order to complete the exercises satisfactorily it is necessary to follow the guidance given in previous chapters. Particular account should be taken of the following.

Chapter 2. General drawing strategy. Drawing construction lines generally. Drawing final lines generally in pencil.

Chapter 3. Using a 1:5 scale. Using a 1:50 scale.

Chapter 4. Size of letters on A4 sheets. Spacing of letters. Annotation. Dimensioning. Horizontal dimensions. Vertical dimensions.

Chapter 5. Lines. Section lines or planes. Representation of materials.

Exercise 6.1 Orthographic projection of a hollow building block

Requirement
Draw the orthographic projection of the hollow building block shown on drawing no. D6/1, in pencil, on a sheet of white A4 paper. You are to draw the plan, longitudinal elevation, end elevation, section A—A and section B—B, but not the three dimensional picture in the bottom right hand corner of drawing no. D6/1. The scale is to be 1:5.

Preliminaries
Begin by drawing a 10 mm border around the edge of the A4 paper. Add the standard title panel at the bottom of the drawing. Remind yourself how to use a 1:5 scale by checking the dimensions on drawing no. D6/1.

Construction lines
Remember that you should first draw all the construction lines very faintly—so that you can just see them. The exact length of these lines is not important.

Draw the plan outline of the building block—a simple rectangle 440 × 100. Then draw the outlines of the longitudinal elevation and section A—A: two rectangles 440 × 215. Then draw the outlines of the end elevation and section B—B: two rectangles 100 × 215. Your drawing should now consist of the construction lines shown on drawing no. D6/1 (first stage).

Draw the three rectangular holes on the plan—the three 120 × 60 rectangles. Then show the position of the holes on both elevations and both sections, using broken lines on the elevations.

Draw the horizontal and vertical dimension lines. Draw the guide lines for the lettering and dimensions.

Final lines
Line in the drawing starting from the top left hand corner of the sheet and working downwards and across to the bottom right hand corner. Use thin lines to line in the plan, longitudinal elevation and end elevation, including the hidden lines. Use thick lines to line in the solid concrete parts of section A—A and section B—B.

Line in the section lines. Line in the horizontal and vertical dimension lines.

Lettering
Add lettering and dimensions between guide lines. Complete the title panel.

Exercise 6.2 Orthographic projection of stone coping

Requirement
Draw the orthographic projection of the stone coping shown on drawing no. D6/2, in pencil, on a sheet of white A4 paper. You are to draw the plan, longitudinal elevation, end elevation, section A—A and section B—B, but not the three dimensional picture in the bottom right hand corner of drawing no. D6/2. The scale is to be 1:5.

Preliminaries
Undertake the preliminaries as described for Exercise 6.1.

Construction lines
Draw the plan outline of the coping—i.e. a rectangle 330 × 215. Next draw rectangles which will contain the overall shapes of the two elevations and the two sections. The containing rectangles for the longitudinal elevation and section A—A will be 330 × 125, and for the end elevation and section B—B will be 215 × 125. Your drawing should now consist of the continuous construction lines shown on drawing no. D6/2 (first and second stage).

Indicate the hipped ends on the plan of the coping by drawing the four lines at 45 degrees. Indicate the ridge by drawing a horizontal line joining the two hipped ends. Complete the construction lines to the elevations and sections by drawing the broken lines shown on drawing no. D6/2 (first and second stage).

Draw the horizontal and vertical dimension lines. Draw the guide lines for the lettering and dimensions.

Final lines
Line in the drawing starting from the top left hand corner of the sheet and working downwards and across to the bottom right hand corner. Use thick lines for section A—A and section B—B, and thin lines elsewhere.

Line in the section lines. Hatch the sections A—A and B—B to indicate stonework. Line in the horizontal and vertical dimension lines.

Lettering
Add lettering and dimensions between guide lines. Complete the title panel.

Exercise 6.3 Orthographic projection of a garden store

Requirement
Draw the orthographic projection of the garden store shown on drawing no. D6/3, in pencil, on a sheet of white A4 paper. You are to draw the plan, roof plan, back, front and side elevations, and sections A—A and B—B, but not the three dimensional picture in the bottom right hand corner of drawing no. D6/3. The scale is to be 1:50.

Preliminaries
Undertake the preliminaries as described for Exercise 6.1, except that you are to remind yourself how to use a 1:50 scale.

Construction lines
Draw the two plan outlines of the store. Ignore the roof overhang at this stage, and draw two rectangles 1220 wide × 1340 deep. Draw rectangles defining the elevations and sections from the underside of the concrete floor slab to the underside of the roof slab. This means that the front and back elevations and section B—B will each be rectangles 1220 wide × 2335 high, and the side elevation and section A—A will be rectangles 1340 wide × 2335 high. Your drawing should now consist of the continuous construction lines shown on drawing no. D6/3 (first and second stage).

Draw the roof slab on the roof plan, and on the sections and elevations. Draw the floor slab and hardcore on the sections. Draw the wall thicknesses on the plan and sections. The above items are all indicated by broken lines on drawing no. D6/3 (first and second stage).

Draw the lines indicating the door on the section A—A, the front elevation and the plan. Draw the door swing on the plan. Draw the horizontal and vertical dimension lines. Draw the section lines. Draw the guide lines for the lettering and dimensions.

Final lines
Line in the drawing starting from the top left hand corner of the sheet and working downwards and across to the bottom right hand corner. Use thick lines to define the walls on plan and sections, the reinforced concrete roof slab on sections, and the concrete floor slab on sections. Thin lines should be used elsewhere. Line in the section lines. Line in the horizontal and vertical dimension lines.

Lettering
Add lettering and dimensions between guide lines. Complete the title panel.

Self Assessment Guide

In order to satisfy yourself that you have completed the exercises to an acceptable standard, study your drawings and check that you can answer YES to the following questions.

Exercise 6.1

1 Does your drawing include all the views and generally look the same as drawing no. D6/1 except that the three dimensional picture in the bottom right hand corner has been omitted?
2 Is your drawing accurately drawn to a scale of 1:5?
3 Have thick lines been used to define the concrete areas seen in section A—A and section B—B, with thin lines used in all other positions?

Exercise 6.2

1 Does your drawing include all the views and generally look the same as drawing no. D6/2 except that the three dimensional picture in the bottom right hand corner has been omitted?
2 Is your drawing accurately drawn to a scale of 1:5?
3 Have thick lines been used to define section A—A and section B—B, with thin lines used in all other positions?

Exercise 6.3

1 Does your drawing include all the views and generally look the same as drawing no. D6/3 except that the three dimensional picture in the bottom right hand corner has been omitted?
2 Is your drawing accurately drawn to a scale of 1:50?
3 Have thick lines been used to define the faces of the walls on plan and section, and the roof slab and floor slab on section, with thin lines used in all other positions?

Now check, with all your drawings, that you can also answer YES to all the questions listed in the 'Standard Self Assessment Questions' at the back of the book (page 48).

7

Isometric and Axonometric Projection

Isometric Projection

Isometric projection is a method of showing three faces of an object on one drawing. The word *isometric* means 'equal measure' and the basis of isometric projection is that three lines defining the three faces or planes produce three equal angles of 120 degrees. This is illustrated in drawing no. D7/1, which shows, in isometric projection, a solid concrete building block and a hollow concrete building block. Isometric projection is achieved in practice by drawing all vertical lines as vertical, and all horizontal lines at 30 degrees to the horizontal, sloping either to the right or left.

It is normal practice to make the measurements on isometric drawings to the same scale as used for plans, elevations and sections. This is not mathematically correct however, and slightly distorts the appearance of the plan view. It is possible to produce an accurate isometric drawing by using a special scale, but consideration of this aspect is inappropriate to an introductory book on construction drawing.

Further examples of isometric projection are given on drawing no. D7/2, which illustrates a coping stone and a garden store.

Axonometric Projection

Axonometric projection is another way of showing objects in a three dimensional form. It is similar to isometric projection except that the horizontal lines are generally drawn at 45 degrees to the horizontal.

The advantage of this method is that the plan is shown as a true plan. This makes it particularly useful for illustrating interior views of buildings, (see drawing no. D7/4) and showing service ducts and pipe runs (see Fig. 7.1).

The principle of axonometric projection is illustrated in drawing no. D7/3 which shows the concrete building blocks in axonometric projection. In practice it is best to begin by drawing the plan—this is a true plan but turned to an angle of 45 degrees. The vertical lines can then be drawn vertically downwards and upwards from the plan. All lines which would be horizontal on the plan view of an orthographic projection will be drawn at 45 degrees to the horizontal sloping either to the left or right on axonometric projection.

A further example of axonometric projection is given on drawing no. D7/4 which shows the interior of a small bungalow.

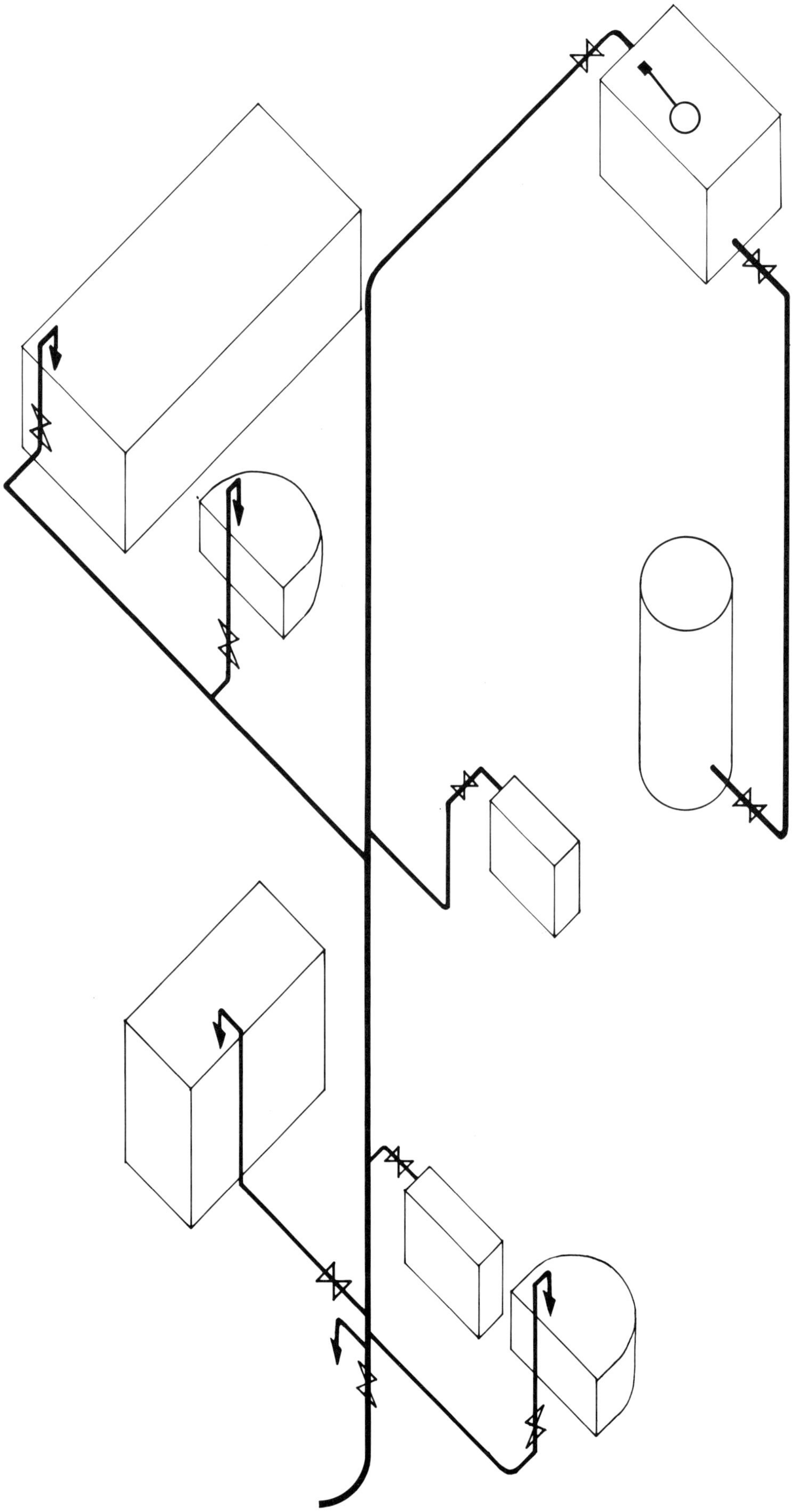

Fig. 7.1 Axonometric of water service (Gold Supply)

Exercise 7.1 Isometric projection of building blocks

Requirement
Draw the isometric projection of the building blocks shown on drawing no. D7/1, in pencil, on a sheet of white A4 paper. The scale is to be 1:5.

Preliminaries
Draw a 10 mm border around the edge of the A4 paper. Add the standard title panel at the bottom of the drawing.

Construction lines
As was mentioned at the beginning of this chapter, the basis of isometric projection is that three lines which meet together each enclose an angle of 120 degrees. The meeting point for the solid building block shown on the top half of drawing no. D7/1 is marked as X. Select a similar point X on your sheet. From X draw a vertical line in a downward direction, and mark off a dimension of 215. Then from X draw a line 30 degrees upwards to the left, and mark off a dimension of 440. Then from X draw a line 30 degrees upwards to the right, and mark off a dimension of 100. Complete the shape of the solid concrete building block by drawing the remaining vertical lines and the 30 degrees lines shown on drawing no. D7/1.

Draw the shape of the hollow concrete building block in a similar fashion. Draw the three rectangular holes on top of the block—i.e. the three 120 × 60 rectangles. Draw the three short vertical lines in these holes.

Draw the slanting and vertical dimension lines. Draw the guide lines for the lettering and dimensions.

Final lines
Line in the drawing starting from the top left hand corner of the sheet and working downwards and across to the bottom right hand corner. All lines are to be thin lines.

Line in the dimension lines.

Lettering
Add lettering and dimensions between guide lines. Complete the title panel.

Exercise 7.2 Isometric projection of a stone coping

Requirement
Draw the isometric projection of the stone coping shown on drawing no. D6/2, in pencil, on a sheet of white A4 paper. The scale is to be 1:5.

Preliminaries
Undertake the preliminaries as described previously.

In order to complete the exercise you will need to study both drawing numbers D6/2 and D7/2.

Your drawing will show the coping only, and this should be positioned centrally on the sheet.

Construction lines
First draw the bottom part of the coping. This is a cuboid, 330 long × 215 wide × 75 high. Draw it in the same way that you drew the solid building block in exercise 7.1. This is illustrated by the solid lines of the

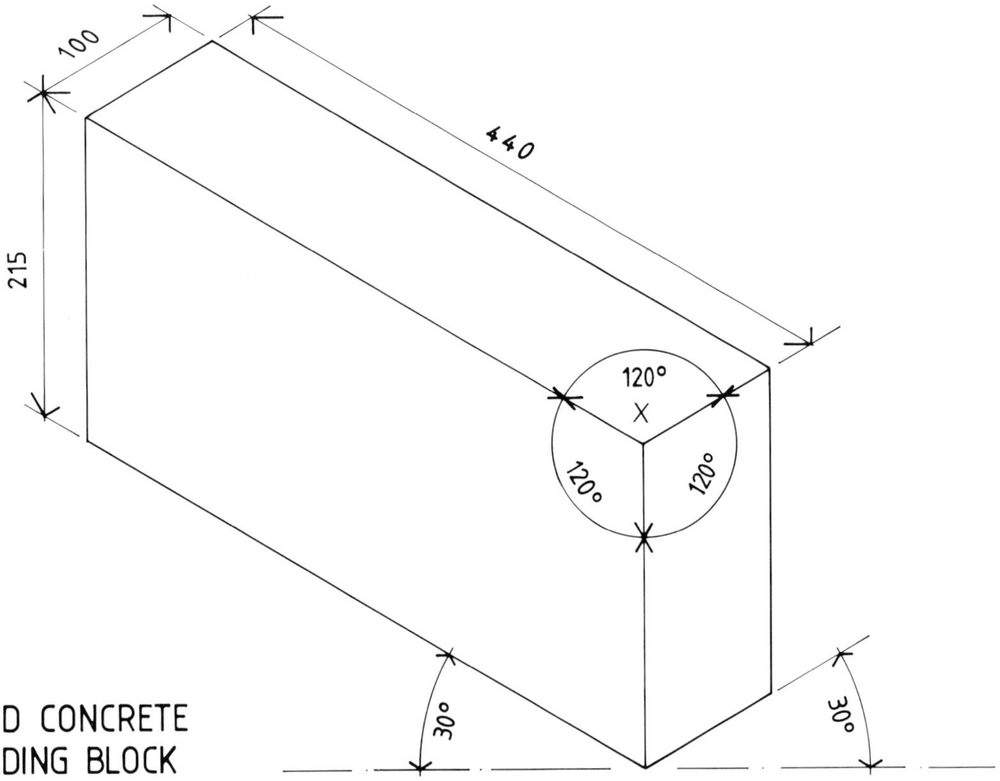

SOLID CONCRETE
BUILDING BLOCK

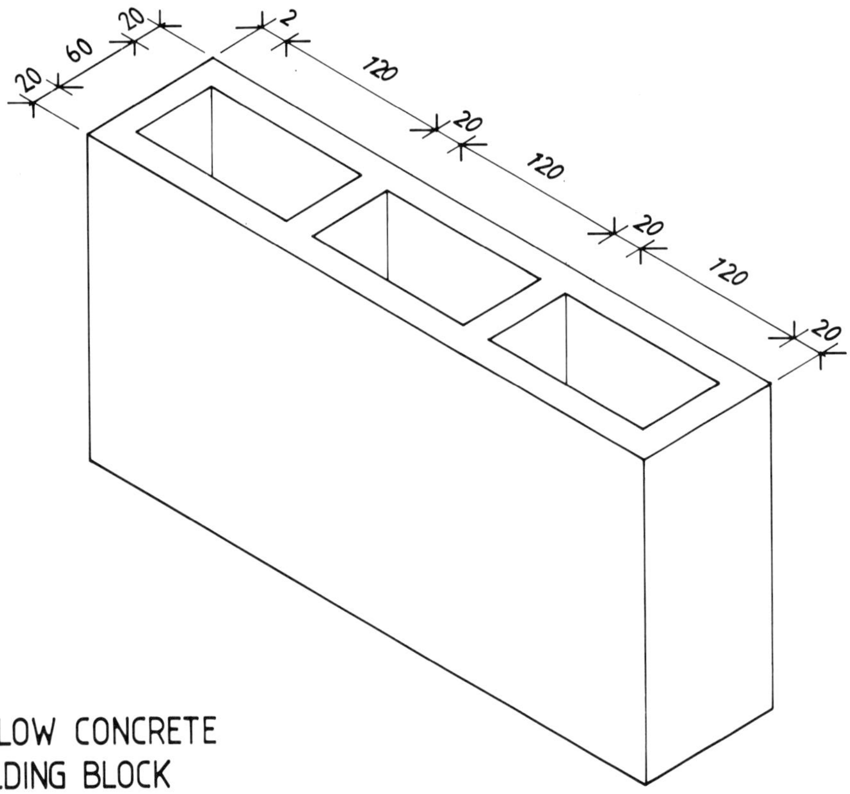

HOLLOW CONCRETE
BUILDING BLOCK

ISOMETRIC PROJECTION OF BUILDING BLOCKS	NAME DATE SCALE 1:5	D7/1

STONE COPING

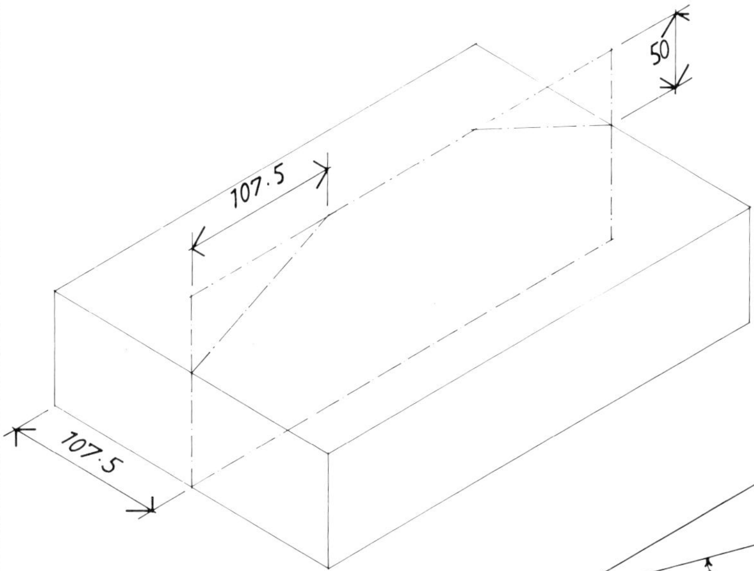

1st STAGE

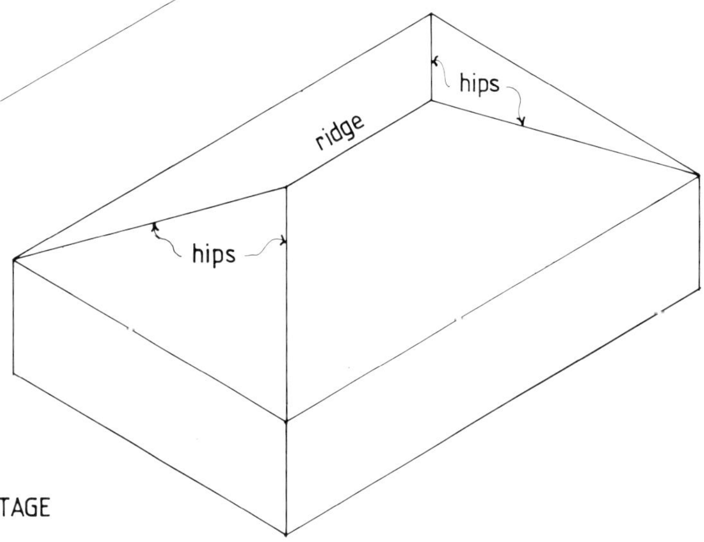

FINAL STAGE

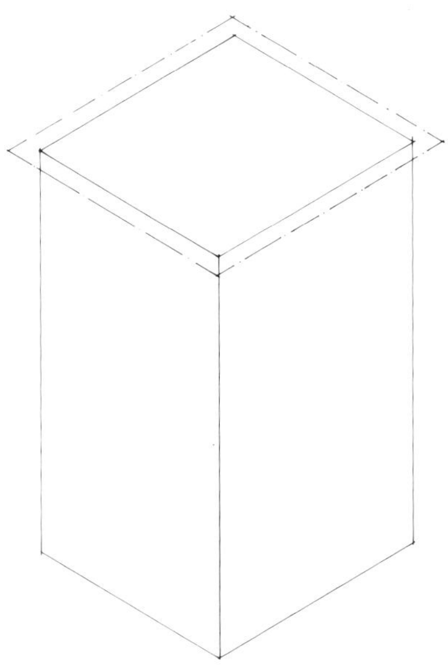

1st STAGE

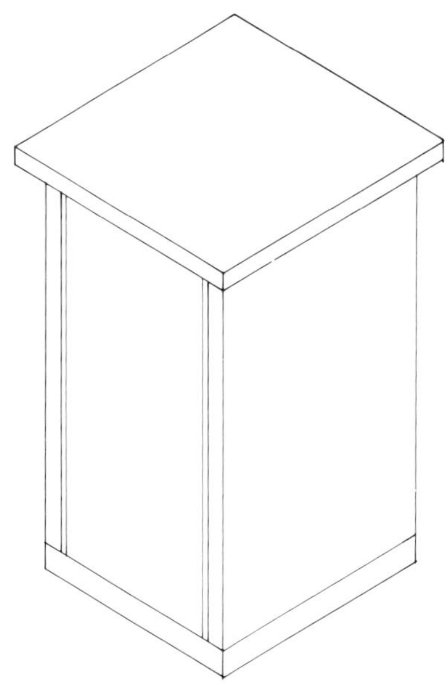

FINAL STAGE

ISOMETRIC PROJECTION OF COPING and GARDEN STORE	NAME	D7/2
	DATE	
	SCALES 1:5 1:50	

isometric in the top left hand corner of drawing no. D7/2. You will see that on this same illustration the longitudinal section A–A is indicated by chain dotted lines. Add this section to your isometric. You have now defined the position of the ridge, and can complete the construction lines for the coping by drawing the four hip lines (see illustration subtitled 'final stage').

Draw the guide lines for the lettering.

Final lines
Line in the drawing in the usual way using thin lines throughout.

Lettering
Add lettering between guide lines. Complete the title panel.

Exercise 7.3 Isometric projection of a garden store

Requirement
Draw the isometric projection of the garden store shown on drawing no. D6/3, in pencil, on a sheet of white A4 paper. The scale is to be 1:20–NOT 1:50.

Preliminaries
Undertake the preliminaries as described previously.

In order to complete the exercise you will need to study both drawing numbers D6/3 and D7/2.

Familiarize yourself with a 1:20 scale.

Construction lines
First draw the basic shape of the store. This is a cuboid, 1220 wide × 1340 deep × 2335 high. Draw it in the same way that you drew the solid building block in exercise 7.1. This is illustrated by the solid lines of the isometric in the bottom left hand corner of drawing no. D7/2.

Now draw the shape of the soffit of the reinforced concrete roof slab. This is indicated by the broken lines of the isometric previously referred to. Now complete the construction lines for the roof slab. This is a shallow cuboid 1420 wide × 1540 deep × 100 high.

Complete the construction lines for the garden store by adding lines defining the top of the concrete floor slab and the door and door frame.

Final lines
Line in the drawing in the usual way using thin lines throughout.

Lettering
Add lettering between guide lines to the title panel.

Exercise 7.4 Axonometric projection of building blocks

Requirement
Draw the axonometric projection of the building blocks shown on drawing no. D7/3, in pencil, on a sheet of white A4 paper. The scale is to be 1:5.

Preliminaries
Undertake the preliminaries as described previously.

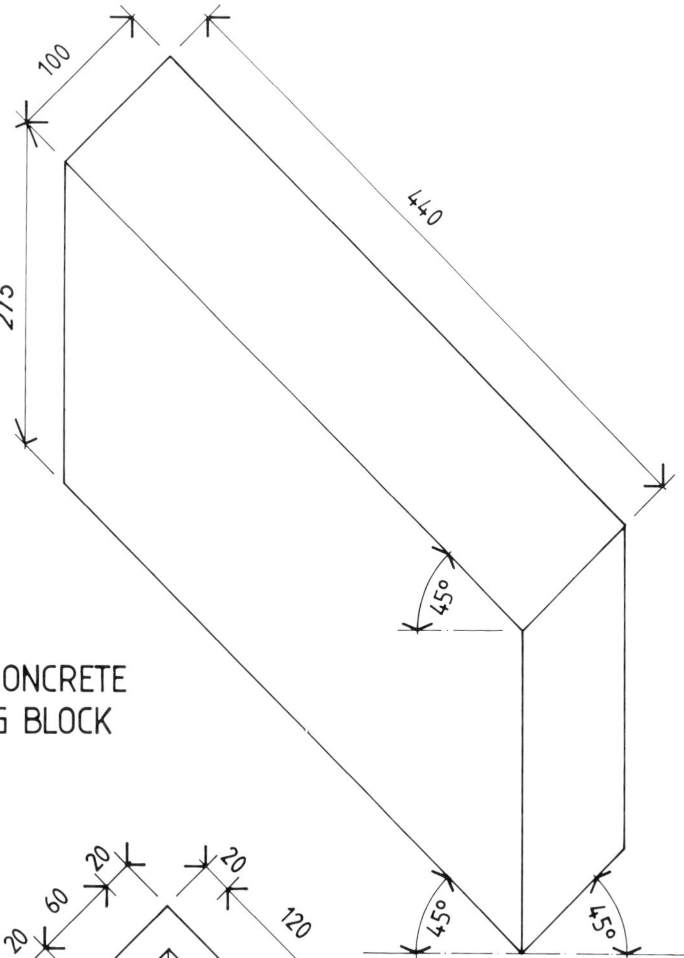

SOLID CONCRETE
BUILDING BLOCK

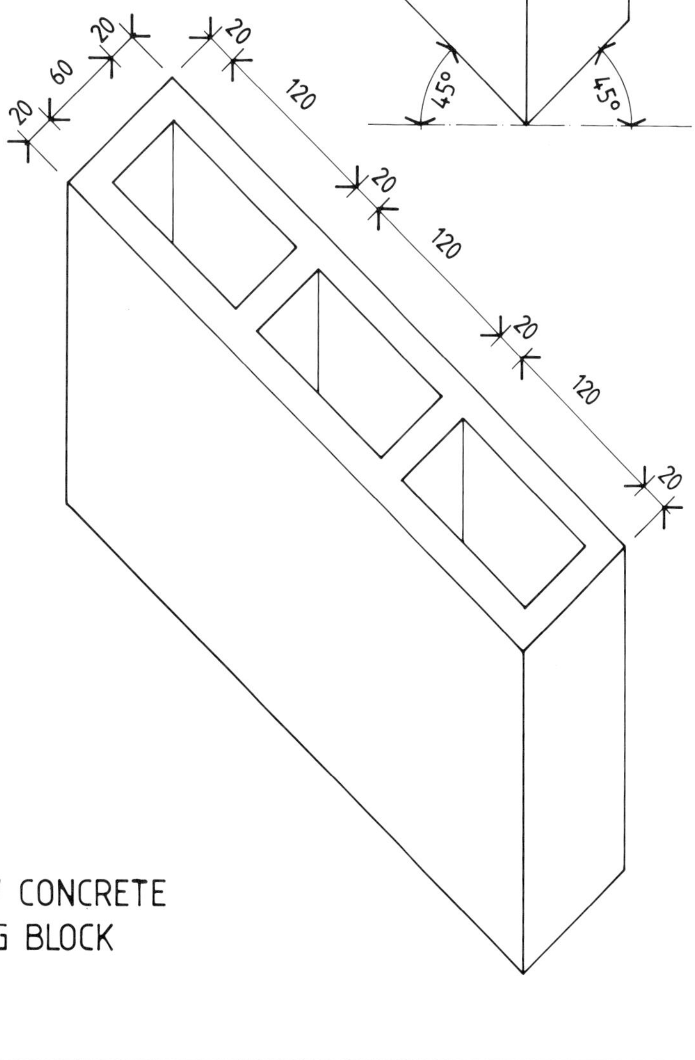

HOLLOW CONCRETE
BUILDING BLOCK

AXONOMETRIC PROJECTION OF BUILDING BLOCKS	NAME DATE SCALE 1:5	D7/3

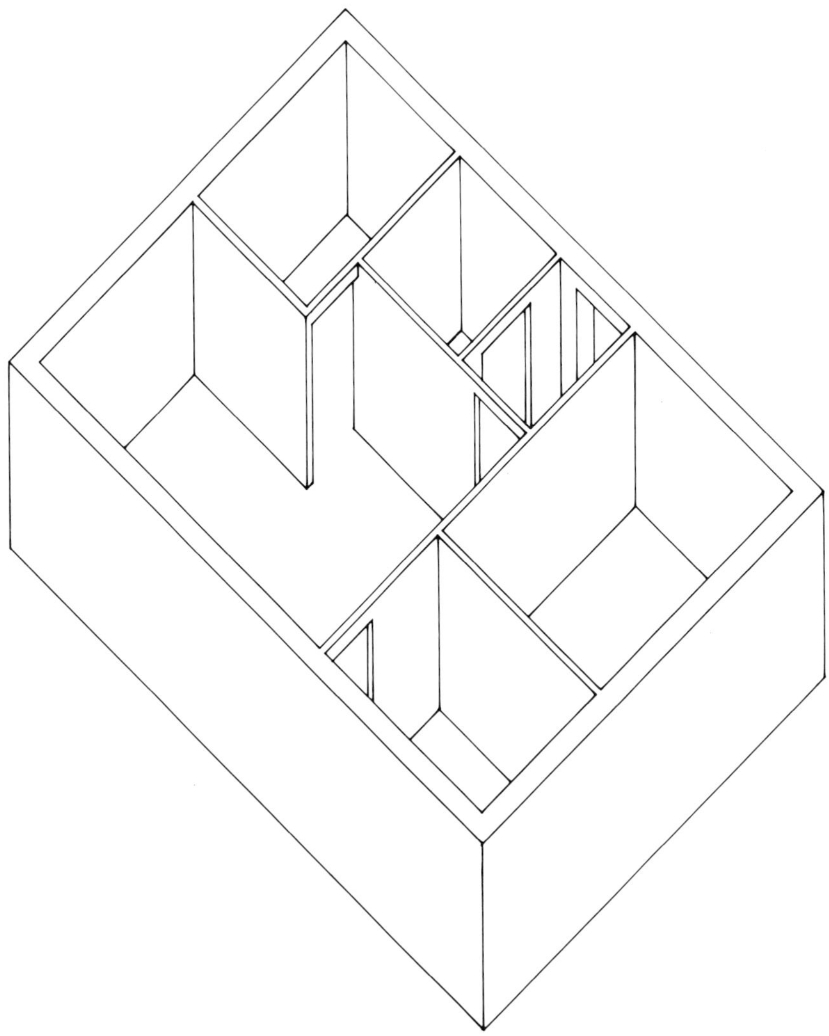

AXONOMETRIC PROJECTION OF BUNGALOW INTERIOR	NAME	D7/4
	DATE	
	SCALE 1:100	

Construction lines
First draw the plan of the solid building block, but set the sides at an angle of 45 degrees to the horizontal. This will be a rectangle 440 × 100.

Now draw the three vertical edges of the building block and mark off a distance of 215 along each edge. Join these three marks to complete the shape of the building block.

Draw the outline shape of the hollow concrete building block in a similar fashion to the solid block. Add the three holes—three rectangles 120 × 60. Draw the three short vertical lines in these holes.

Draw the slanting and vertical dimension lines. Draw the guide lines for the lettering and dimensions.

Final lines
Line in the drawing in the usual way using thin lines throughout. Line in the dimension lines.

Lettering
Add lettering and dimensions between guide lines. Complete the title panel.

Exercise 7.5

Axonometric projection of bungalow interior

Requirement
Draw a copy of the axonometric projection of the bungalow interior on drawing no. D7/4, in pencil, on a sheet of white A4 paper. The scale is to be 1 : 100.

Construction lines
First copy the outline plan of the bungalow. This is a rectangle 9100 long × 6600 wide, set at an angle of 45 degrees to the horizontal. Then draw the inner faces of the external walls, making the walls 300 thick.

Copy the arrangement of the internal walls by scaling all the dimensions from drawing no. D7/4 and transferring these dimensions to your own drawing.

Draw the vertical lines defining the height of external walls and internal partitions.

Draw the 45 degree lines defining the lines of the bottom of the walls and partitions.

Draw the vertical and 45 degree lines defining the doors.

Final lines
Line in the drawing in the usual way using thin lines throughout.

Lettering
Add lettering between guide lines to the title panel.

Self Assessment Guide

In order to satisfy yourself that you have completed the exercises to an acceptable standard, study your drawings and check that you can answer YES to the following questions.

Exercise 7.1

1 Does your drawing include both the solid and the hollow concrete building blocks and generally look the same as drawing no. D7/1?
2 Is your drawing accurately drawn to a scale of 1:5?
3 Have you used thin lines throughout?

Exercise 7.2

1 Does your isometric projection generally look the same as the one near the top right hand corner of drawing no. D7/2?
2 Is your drawing accurately drawn to a scale of 1:5, strictly in accordance with the dimensions given on drawing no. D6/2?
3 Have you used thin lines throughout?

Exercise 7.3

1 Does your isometric projection generally look the same as the one near the bottom right hand corner of drawing no. D7/2, except that it is drawn to a larger scale?
2 Is your drawing accurately drawn to a scale of 1:20, strictly in accordance with the dimensions on drawing no. D6/3?
3 Have you used thin lines throughout?

Exercise 7.4

1 Does your drawing include both the solid and the hollow concrete building blocks, and generally look that same as drawing no. D7/3?
2 Is your drawing accurately drawn to a scale of 1:5?
3 Have you used thin lines throughout?

Exercise 7.5

1 Does your axonometric projection look the same as the one on drawing no. D7/4?
2 Is your drawing accurately drawn to a scale of 1:100, as drawing no. D7/4?
3 Are all the vertical lines exactly vertical, and all the inclined lines drawn at exactly 45 degrees?
4 Have you used thin lines throughout?

Now check, with all your drawings, that you can also answer YES to all the questions listed in the 'Standard Self Assessment Questions' at the back of the book (page 48).

8

Types of Construction Drawings

Introduction

There is a wide range of different types of drawings prepared for construction purposes, many of them by specialists. In order to set in context the remaining work the reader will be doing in this introductory career stage, a summary is given in this chapter of the various kinds of construction drawings.

The broad principles of draughting techniques apply to all the different types described below, but details, such as graphic conventions, vary for each discipline.

Planning and Survey Drawings

The first category of drawings affecting the construction process are planning and survey maps, drawings and diagrams, showing existing topography, physical features and building services, together with information as to the type of development permitted in a particular area.

Planning maps and drawings are generally the work of town planners and planning technicians. Structure plans are prepared by the county councils, and local plans by the district councils. Both aim to allocate appropriate areas of land for different types of development—housing, industrial, commercial, recreation etc. They further seek to protect and conserve the natural and built environment.

Planning drawings are often prepared over ordnance survey sheets. Common scales for maps are 1:1 000 000, 1:500 000, 1:200 000, 1:100 000, 1:50 000, and for town surveys 1:50 000, 1:20 000, 1:10 000, 1:5000 and 1:2500.

Survey drawings are prepared by land surveyors and surveying technicians to show the shape, contours and physical features of land, such as boundaries, means of access, fences, trees, hedges, lakes, ponds, rivers and streams, railways, cuttings and embankments. This information is needed before architects can begin work on their preliminary design drawings. Land surveyors, like town planners, often make use of ordnance survey maps when preparing their survey drawings. Common scales for survey drawings are 1:2500, 1:2000, 1:1250, 1:1000 and 1:500.

Architectural Preliminary Sketch Design Drawings

When architects have obtained planning information regarding the type of development likely to be permitted in a particular area, and survey drawings detailing the shape, contours and physical features of the land, they are ready to begin work on their preliminary sketch design drawings.

These preliminary drawings may be drawn freehand and will show only the broad details of a possible scheme, giving general intentions—often in a diagramatic form—and indicating the general allocation of space for various purposes.

Architectural Design Drawings

The design drawings will follow on from the preliminary drawings and may be prepared in a number of stages. The final design drawings will be produced to a high standard as they are presentation drawings explaining, and sometimes attempting to sell, the scheme to the client, planning authorities and others.

They will generally include site plans, floor plans, elevations and perspective views. The architects and architectural technicians involved in the preparation of these drawings will often use the expertise of structural engineers, building services engineers and others when preparing their drawings. Although the drawings will show only a limited amount of constructional and other detail, it is obviously essential that they represent proposals which can be translated into a practical building solution.

Architectural Production Drawings

When the final design drawings have been completed and approved the architectural team are ready to prepare their production drawings. These drawings give precise details of the construction of the building, including all fixtures, fittings and finishes.

They will be needed by the local authority building control officers, so they can consider and hopefully approve them under the building regulations; by the structural engineers, building services engineers and other consultants, so they can prepare their own specialist details; by the quantity surveyors, so they can produce bills of quantities which detail the materials and workmanship needed to complete the project; by the contractors tendering for the contract; by the site staff of the selected contractor, so they know how to construct the buildings and other works; and by clients, so they are aware of what they are getting for their money.

It is obvious from the above list that the production drawings must be detailed and explicit. Most of the drawings will be in orthographic projection, but use may be made of three dimensional projection methods such as isometric and axonometric. Some of the drawings, particularly in the case of larger projects, will be prepared in stages, and some will initially be prepared in sketch form.

Architects have the choice of structuring their production drawings in various ways. The most common method is the *types of information* structure, in which the drawings are divided into the following types.

(a) *Location drawings*, commonly consisting of site plans, floor plans, sections and elevations. These drawings show where the work will be, but may not show how the work is to be performed. Other examples of location drawings which will sometimes be required, particularly on larger projects, are block plans, landscaping plans, drainage plans, foundation plans, ceiling plans and roof plans. Information on location drawings is given in Chapter 9 (site plans); Chapter 10 (floor plans); Chapter 11 (sections); and Chapter 12 (elevations).

(b) *Assembly drawings*. These show how the building is to be put together on the site. Information on assembly drawings is given in Chapter 13.

(c) *Component drawings*. These show details of items manufactured on and off site. Information on component drawings is given in Chapter 14.

Structural and Civil Engineering Drawings

On larger construction projects structural engineers and structural engineering technicians are employed to assist in the design, drawing and supervision of the structural elements of the building. They may be involved in a wide range of different construction methods including structural brickwork and timber, but the two commonest methods are structural steelwork and reinforced concrete. Both of these techniques involve the preparation of general arrangement plans, elevations and sections drawn to scales such as 1:100, and details drawn to scales such as 1:10 and 1:20.

On some large projects civil engineering elements may also be involved, such as roads, bridges, tunnels and similar unroofed structures, and drawings will be required of these items.

Structural and civil engineering drawings are dealt with in Chapter 15.

Building Engineering Services Drawings

Another group of specialists employed on larger projects are building services engineers and technicians. They will be concerned with the design, drawing, construction and supervision of the building services. These are generally assumed to include hot and cold water supplies, above and below ground drainage including sanitary appliances, refuse disposal, heating, ventilation, air conditioning, electrical installations including lighting, telecommunications, gas installations, fire protection, mechanical conveyors and security systems.

Building services engineers and technicians will use some of the architectural location drawings—e.g. the 1:100 plans and sections—to show the general arrangements for the building engineering services. Generally they will add their information to copy negatives—copies made on tracing paper or tracing film—of the architectural drawings, copied before the detailed constructional information has been added to the drawings.

As a rule only one service will be shown on each drawing. In addition to the general arrangement drawings they will prepare detailed drawings to a larger scale of key areas of the building engineering services; for example, 1:20 and 1:50 plans and sectional elevations of boiler rooms, air handling plant rooms and electrical sub-stations.

On smaller jobs where the building services are relatively uncomplicated the drawings for this part of the project will normally be done by the architectural staff.

Building engineering services are dealt with in Chapter 16.

Subcontractors' and Suppliers' Drawings

Most construction contracts involve the use of subcontractors and suppliers. They are responsible for specific parts of the construction work, and for supplying materials used in the construction. Examples are the supply and the supply and fixing of steel windows, colour coated cladding for industrial buildings, partitions, suspended ceilings, shop fronts, escalators and stonework. Subcontractors and suppliers are often responsible for providing the drawings covering their own area of specialist work.

Contractors' Drawings

The contractor may need to provide drawings for temporary works items. This is work required to facilitate the construction process, but which will not remain in the finished building. Examples are plans showing the layout of the temporary huts, services, storage areas, means of access etc. needed during the building operations. Other examples include timbering—temporary supports to the sides of excavations; and formwork—temporary supports to in situ reinforced concrete work.

Measured Drawings

These are drawings prepared by architects and architectural technicians to illustrate existing buildings. They are used when alterations and extensions are required to buildings, or sometimes as a requirement under town planning legislation in connection with the preservation of buildings of historical interest.

Measured drawings are so called because they are based on carefully taken measurements of existing buildings. The type of drawing ranges from small scale plans and elevations to large scale details. Considerable care and accuracy is required in producing this type of drawing.

Record Drawings

Record drawings are drawings prepared at the close of a construction contract to record exact details of the completed building. This means ensuring that drawings are updated to show changes made during the construction process. It is particularly important to record items which will be hidden on completion of the work—for example the precise position of underground drains.

Freehand Drawings

An important group of construction drawings are those drawn freehand—i.e. without the help of set-squares, tee-squares and other aids. They are used by all members of the design team, but particularly by architects. Sometimes they are referred to as sketch drawings or freehand sketches.

Students need to be able to draw freehand sketches and diagrams for inclusion in their notes and assignments, and so the ability to sketch quickly and accurately is useful from the very beginning of their career in the construction industry.

Freehand drawings are discussed further in Chapter 17.

Self Assessment Guide

In order to check whether you have understood all you have read in this chapter, you should now attempt to answer the following questions.

1. What information is shown on planning maps and drawings?
2. What information is shown on survey drawings?
3. What types of drawings are normally included in the architectural final design drawings?
4. What is the purpose of architectural production drawings?
5. What are the two commonest structural systems designed by structural engineers?
6. What information is included on drawings produced by building services engineers?
7. What drawings are the responsibility of subcontractors and suppliers?
8. What drawings are prepared by the contractor?
9. What is meant by the term *measured drawings*?
10. What is meant by the term *record drawings*?
11. Who prepares and uses freehand drawings?

When you have attempted all the above questions, reread the appropriate part of the chapter and check you know the correct answer to every question.

9

Site Plans

Introduction

A site plan is a location drawing, and like most plans is a view looking downwards. It supplies a bird's eye view of the shape, size and layout of the entire site. An example of a site plan is shown on drawing no. D9/1.

Purpose

The purpose of a site plan is to

(a) provide a general picture of the site, including its shape and extent;

(b) locate the buildings and other elements of the project—e.g. roads, garden walls and landscaping—both horizontally and vertically;

(c) indicate the levels and surface features of the finished site;

(d) sometimes provide information on external services, especially underground drainage.

Scale

Where possible it is best to use a site plan with a scale of 1:200. This scale is particularly suitable for smaller sites and allows the maximum amount of detail to be shown. However in order to avoid over-large drawings it is often necessary to draw site plans for larger sites to a scale of 1:500.

Dimensions

It is important that site plans, like other drawings, be adequately dimensioned. Setting out dimensions for the buildings and other physical features, and relating them to fixed points such as an existing road or building is essential. Site boundaries also need to be defined.

Use the same type of dimension lines for all dimensions. Remember to place all dimensions just above the dimension lines so that they can be read from the bottom and right hand side of the drawing. Use guide lines for all dimensions. There is no need to add the suffix 'mm' after the figures.

Datum

As well as positioning the buildings and other features on the site laterally, it is also essential to relate them vertically to a datum. This must be a clearly defined point which will remain undisturbed for the duration of the contract.

Landscaping

Trees, shrubs, flower beds and other landscaping items also need to be shown on site plans. Either spot levels or contour lines will be provided to define the topography of the finished site. If the landscaping is extensive, separate landscape drawings will be provided.

SITE PLAN	NAME	D9/1 1st STAGE
	DATE	
	SCALE 1:200	

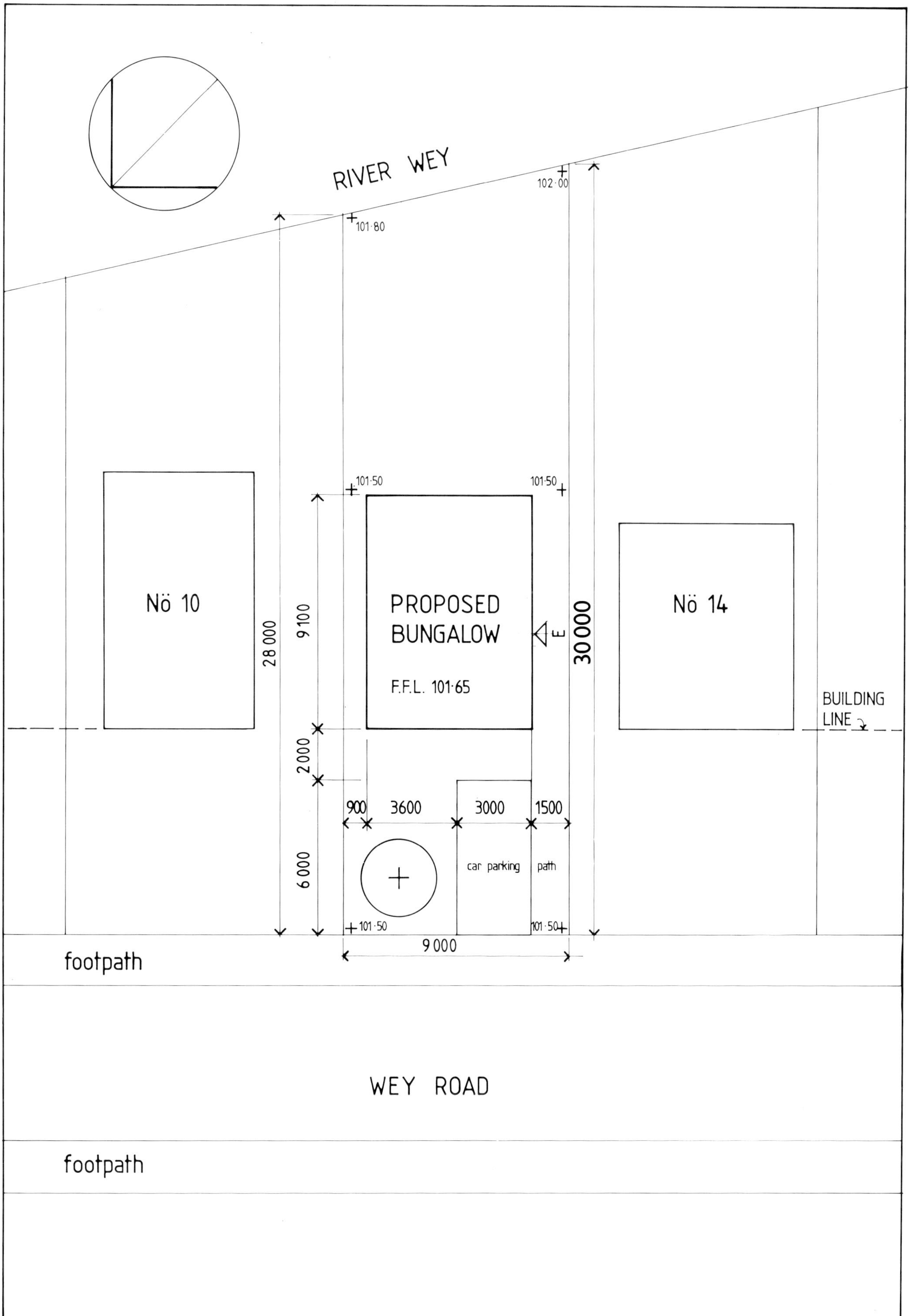

SITE PLAN

NAME

DATE

SCALE 1:200

D9/1

In the case of the site plan shown on drawing no. D9/1, the only landscaping items are a tree in the front garden; a paved area for car parking; and a path to the side of the bungalow.

Services

On small simple projects outline information about external services, particularly underground drainage, may be shown on the site plan. In other circumstances this information will be given on separate services drawings.

North Point

It is important to include a north point so that in verbal and written communication reference can be made to specific locations. For instance: 'the centre line of the surface water drain is to be positioned 1000 mm from the external south west face of the house'.

Exercise 9.1

Site plan

Requirement
Draw the site plan shown on drawing no. D9/1, in pencil, on a sheet of white A4 paper. The scale is to be 1:200.

Preliminaries
Undertake the preliminaries as described for previous exercises.

Construction lines
Remember that you should first draw all the construction lines very faintly—so that you can just see them. As previously mentioned the exact length of these construction lines is not important.

Because the top (south east) boundary of the site is not at right angles to the side boundaries it will be best to commence drawing from the bottom of the sheet. Carefully read each dimension from drawing no. D9/1 and transfer them to your own drawing with the aid of a 1:200 scale. Any dimensions which are not given can be scaled from drawing no. D9/1.

A suitable order for drawing the various lines is given below.

1 Draw the line representing the outside edge of the bottom path to Wey road; then the bottom (north west) edge of the road; then the top (south east) edge of the road; and finally the front (north west) boundary of the site.
2 Draw the lines representing the left hand side (north east) boundary, and the right hand side (south west) boundary. Measure the 28 000 and 30 000 dimensions along these two lines.
3 Draw the back (south east) boundary and extend this line the full width of your sheet.
4 Draw the horizontal line representing the building line; that is, a distance of 8000 from the front boundary of the site.
5 Draw the back face of the bungalow by measuring a distance of 9100 from the building line. Extend the line the full 9000 width of the site.
6 Draw the left hand (north east) face and the right hand (south west) face of the bungalow.
7 Draw the vertical line denoting the 1500 wide path, and the vertical and horizontal lines denoting the 6000 × 3000 car parking area.
8 By scaling off the distances from drawing no. D9/1, draw the buildings either side of the site—i.e. numbers 10 and 14 Wey Road.

At this stage your drawing should look as indicated on drawing no. D9/1 (first stage).

9 Draw the + signs denoting levels on plans.
10 Draw the circle of the north point, and the circle of the proposed tree. Draw the vertical, horizontal and slanting lines to complete the north point and proposed tree.
11 Draw the horizontal dimension lines, starting at the top and working down the sheet, and the vertical dimensions starting at the left hand side and working towards the right.
12 Draw the guide lines for the lettering and dimensions.

Final lines
Line in the drawing starting from the top left hand corner of the sheet and working downwards and across to the bottom right hand corner. Use thick lines for the four faces of the proposed bungalow and the arrow head of the north point. Use thin lines for all other lines including dimension lines.

Lettering
Add lettering and dimensions between guide lines. Complete the title panel.

Exercise 9.2 Practising tracing in ink

Trace in ink, on a sheet of A4 paper the drawing produced for Exercise 9.1

Self Assessment Guide

In order to satisfy yourself that you have completed the exercises to an acceptable standard, study your drawings and check that you can answer YES to the following questions.

1 Do your drawings look the same as drawing no. D9/1 and include exactly the same information?
2 Are your drawings accurately drawn to a scale of 1:200?
3 Have thick lines been used to define the four faces of the proposed bungalow and the north point arrow, with thin lines used in all other positions?

Now check with your drawings that you can also answer YES to all the questions listed in the 'Standard Self Assessment Questions' at the back of the book (page 48).

10

Location Floor Plans

Introduction

Floor plans are generally the most useful, and the most used of the location drawings. They are really sectional plans because they show the view obtained by cutting horizontally through a building at some point above the floor level. It is assumed that you move away the top part of the building and look down at the plan of the remaining bottom part. This plan view will not only illustrate the arrangement of the rooms and spaces and their shapes, but will also show the thickness of all the external and internal walls.

The level at which you cut horizontally through a building is commonly assumed to be 1 metre above the floor level. This has the advantage of passing through most of the windows and doors, which means that dimensions giving the positions of all openings can be given.

An example of a floor plan is shown on drawing no. D10/1.

Purpose

The purpose of a location floor plan is to

- indicate the shape and layout of the building;
- provide the setting out dimensions for the building;
- locate spaces such as rooms, and parts such as doors;
- provide references stating where more detailed information can be found.

Scale

The most common scale is probably 1:100, but 1:50 is used on small buildings, and 1:200 on large projects. Scales of 1:20 may be used, particularly for enlarging complex areas of larger buildings, such as reception and toilet areas of office blocks and factories.

Dimensions

It is important to dimension floor plans fully, including lengths of all walls or parts of walls, width of all window and door openings, and their lateral positions in the walls.

Remember the information given in previous chapters that dimensioning methods should be standardized, and that figures should be placed above and in the same direction as the dimension lines so they can be read from the bottom and right hand side of the drawing.

Walls and Partitions

Thick lines should be used to define the inside and outside faces of external walls, and both faces of the internal partitions.

Where cavity walls form part of the construction the cavity may be indicated by thin lines, but it is suggested you omit this detail on the 1:50 floor plan you are drawing as an exercise for this chapter.

Hatching is often used, particularly on larger scale plans (see Chapter 5), but can be omitted at this stage. Where floor plans show existing walls, they are often filled in solid.

Windows

As was stated under 'Dimensions' windows will be positioned laterally on the floor plans. Their positions within the wall thickness will be shown on the assembly drawings if these are provided. However, where the scale of the floor plan is 1:50 or larger, it is sensible to locate the windows in approximately their correct positions relative to the wall faces.

It is usual practice to number each window—W1, W2, W3 etc.

Doors

Doors should also be numbered—D1, D2, D3 etc.

At each door position it should be made clear which way the door is hung. This item was covered in Chapter 5.

Other Items

Sanitary fittings, cupboards and other fittings should be shown in outline on floor plans. It is important however not to repeat information which is given on other drawings, such as assembly and component drawings. If too much information is provided, the drawing will become confusing and difficult to read.

Room Names and Notes

A name should be given to each room or space. On large projects room numbers will also be provided.

Notes should be kept to a minimum, and duplication of information provided on other drawings should be avoided.

Take trouble with your lettering. The most important thing is that the lettering should be easy to read. It should be neat and preferably attractive to look at. Always use guide lines, but make these very faint—barely visible.

North Point

A north point is sometimes provided on floor plans.

Exercise 10.1 Floor Plan

Requirement
Draw the floor plan shown on drawing no. D10/1, in pencil, on a sheet of white A4 paper. The scale is to be 1:50.

Preliminaries
Undertake the preliminaries as described for previous exercises.

Construction lines
Draw a rectangle, 9100 high × 6600 wide, defining the four outer faces of the bungalow.

Draw the four inner faces of the 300 mm thick external walls.

Draw the horizontal lines defining the 100 mm partition between the kitchen and bathroom.

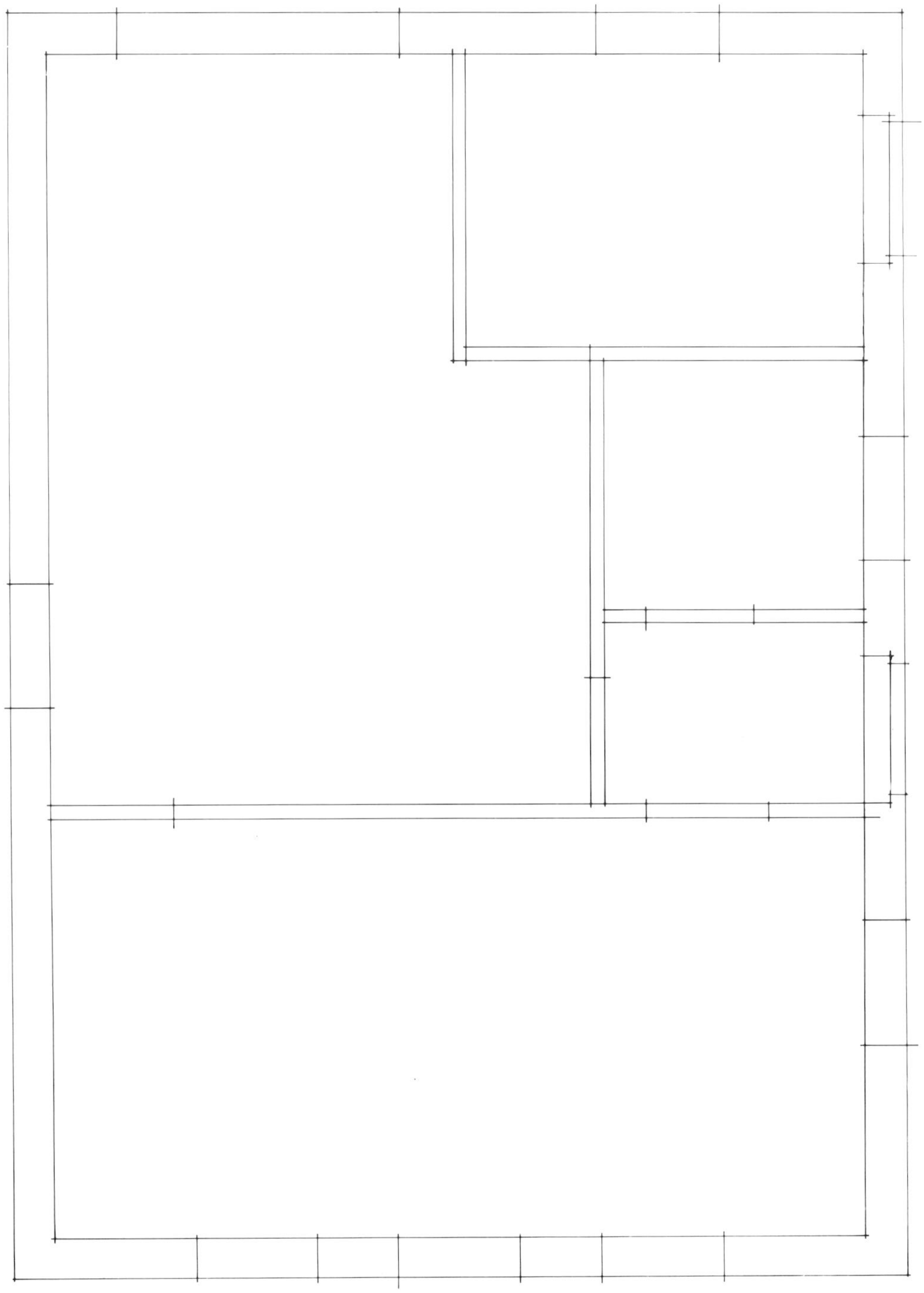

FLOOR PLAN	NAME	D10/1
	DATE	1st STAGE
	SCALE 1:50	

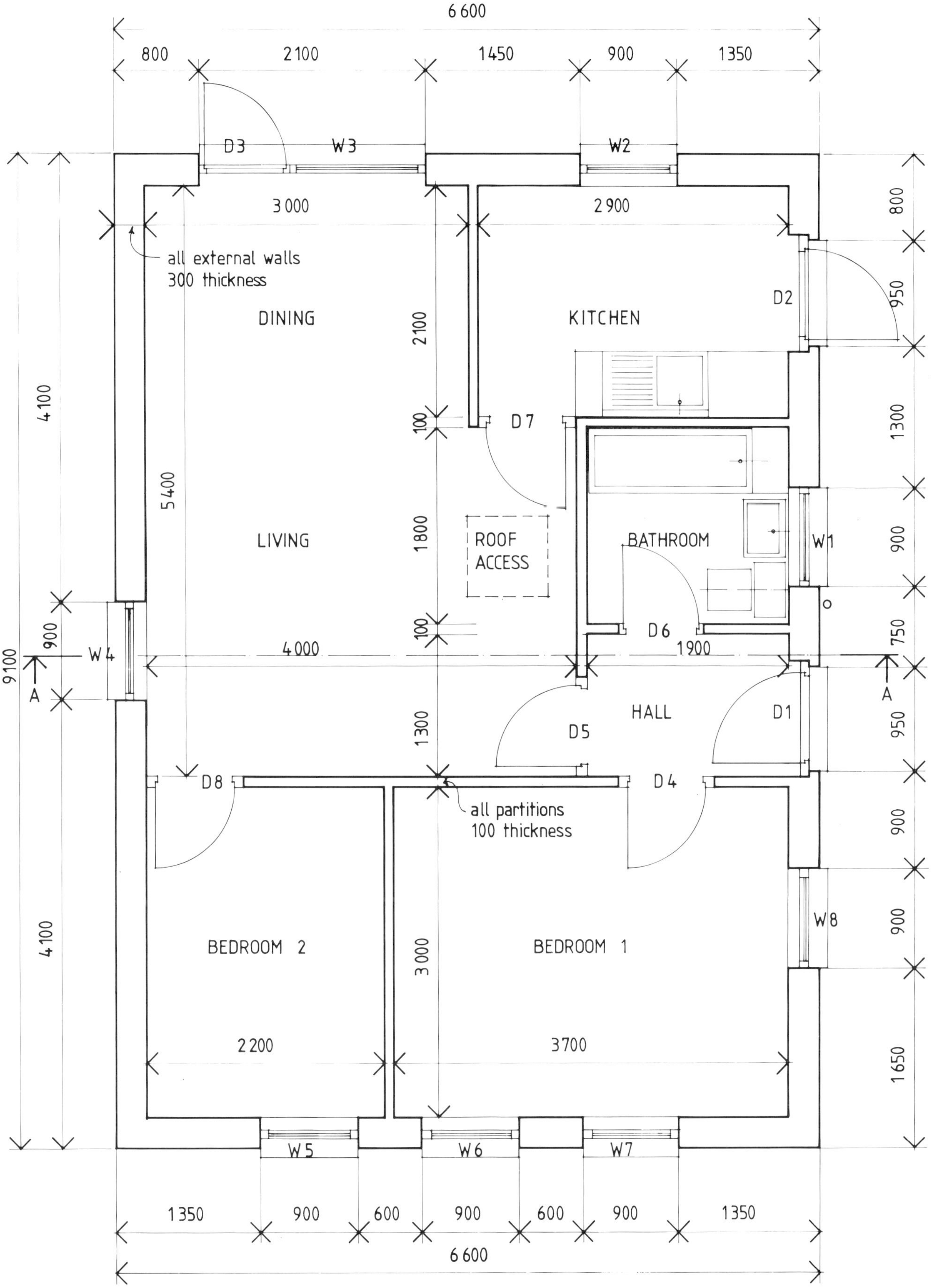

FLOOR PLAN	NAME DATE SCALE 1:50	D 10/1

Draw the horizontal lines defining the 100 mm partition between the bathroom and hall.

Draw the horizontal lines defining the 100 mm partition between the living area/hall and bedroom 2/bedroom 1.

Draw the vertical lines defining the 100 mm partition between the dining area and kitchen.

Draw the vertical lines defining the 100 mm partition between the living area and bathroom/hall.

Draw the vertical lines defining the 100 mm partition between bedroom 2 and bedroom 1.

By using the dimensions where they are given on drawing no. D10/1, and by scaling elsewhere, mark the position of the external windows and doors, and the internal doors.

At this stage your drawing should look as indicated on drawing no. D10/1 (first stage).

By transferring the scaled dimensions from drawing no. D10/1 draw the kitchen cupboards, sink unit, cooker, bath, w.c. and duct.

By scaling distances from drawing no. D10/1 and using broken lines, mark the position of the roof access.

By scaling distances from drawing no. D10/1, draw the lines of door frames and windows.

Using a circle guide or compasses draw the door swings, carefully copying them from drawing no. D10/1.

Draw the straight lines denoting the doors when in open positions, carefully copying them from drawing no. D10/1.

Draw the horizontal and vertical dimension lines.

Draw the guide lines for the lettering and dimensions.

Final lines
Line in the drawing starting from the top left hand corner of the sheet and working downwards and across to the bottom right hand corner. Use thick lines for the external and internal faces of all walls and partitions. Use thin lines for all other lines including dimension lines.

Lettering
Add lettering and dimensions between guide lines. Complete the title panel.

Exercise 10.2 Practising tracing in ink

Trace in ink, on a sheet of A4 paper the drawing produced for Exercise 10.1.

Self Assessment Guide

In order to satisfy yourself that you have completed the exercises to an acceptable standard, study your drawings and check that you can answer YES to the following questions.

1 Do your drawings look the same as drawing no. D10/1 and include exactly the same information?
2 Are your drawings accurately drawn to a scale of 1:50?
3 Have thick lines been used to define the wall and partition faces, with thin lines used in all other positions?

Now check with your drawings that you can also answer YES to all the questions listed in the 'Standard Self Assessment Questions' at the back of the book (page 48).

11

Location Sections

Introduction

A section is a view of a building or object obtained by making an imaginary cut through it. The term section is mainly used where the cut is made in a vertical direction, and this is so in the case of location sections.

A vertical section through a building will show details of the construction of the foundations, walls, floors, roof and other parts. The number of sections required of a building will depend on its size and complexity. Generally there will be at least two sections. One of these will be a cross section, across the width of the building. The other will be a longitudinal section, along the length of the building. Sections are intended to help the builder construct the building, so the exact position of the section should be chosen to show as much construction as possible.

An example of a location section is shown on drawing no. D11/1.

Purpose

The purpose of a location section is to (a) give a vertical view of the building; and (b) provide overall vertical dimensions and levels.

Scale

Generally the scale selected will be the same as used for the location floor plans. A scale of 1:100 is common, but 1:50 is used on small projects, and 1:200 on large projects. Scales of 1:20 or 1:25 may be used, particularly for enlarging complex areas of larger buildings.

Dimensions

The number of dimensions provided will depend on the scale used. The minimum requirement will be floor to floor, or floor to ceiling dimensions, and these will have to be related to the site datum. When the scale is 1:50 or larger, more detail is possible including cill heights and door and window head positions.

Walls and Partitions

Thick lines should be used to define the inside and outside faces of external walls, and both faces of the internal partitions.

Where cavity walls form part of the construction the cavity is often indicated by thin lines.

Hatching is often used, particularly on the larger scale sections.

Other Structural Elements

Thick lines should be used to define both faces of concrete floor and roof slabs.

Thick lines should also be used to define the faces of other structural elements, such as foundations and beams, when these are viewed in section.

Other Elements

Thin lines should always be used to indicate parts of the construction which are seen in elevation—e.g. rafters and ceiling joists. Non structural items—e.g. plasterboard ceilings—should be indicated by thin lines whether seen in section or elevation.

Exercise 11.1 **Section**

Requirement
Draw the section shown on drawing no. D11/1, in pencil, on a sheet of white A4 paper. The scale is to be 1:50.

Preliminaries
Undertake the preliminaries as described for previous exercises.

It is suggested that before you begin drawing, you read all the instructions for this exercise.

Construction lines
First draw four horizontal lines defining the following key levels.

1 Line of the underside of the foundations.
2 Line of the top of the foundations. This is a distance of 300 mm above the previous line.
3 Line of the top of the concrete floor slab (SFL). This is a distance of 900 mm above the previous line.
4 Line of the underside of the ceiling joists. This is a distance of 2340 above the previous line.

Now draw six vertical lines defining important vertical positions as follows.

1 Line of the outside face of the left hand external wall.
2 Line of the inside face of the left hand external wall. This is a distance of 300 mm to the right of the previous line.
3 Line of the left hand face of the internal partition. This is a distance of 4000 mm to the right of the previous line.
4 Line of the right hand face of the internal partition. This is a distance of 100 mm to the right of the previous line.
5 Line of the inside face of the right hand external wall. This is a distance of 1900 mm to the right of the previous line.
6 Line of the outside face of the right hand external wall. This is a distance of 300 mm to the right of the previous line.

Check that the distance between the external wall faces scales 6600 mm, and that the distance between the underside of the foundations and the underside of the ceiling joists scales 3540. Make any necessary adjustments.

At this stage your drawing should look as indicated on drawing no. D11/1 (first stage).

Now look at drawing no. D13/4, which is an assembly drawing of the roof eaves. Draw a line as close as possible below the underside of the ceiling joist to represent the bottom surface of the ceiling finish (see page). Draw the wall plates as shown on drawing D13/4. NB the top of the wall plate is at the same level as the underside of the ceiling joist.

Using an adjustable set-square (or a protractor) draw the inclined lines at an angle of 40 degrees to indicate the rafters.

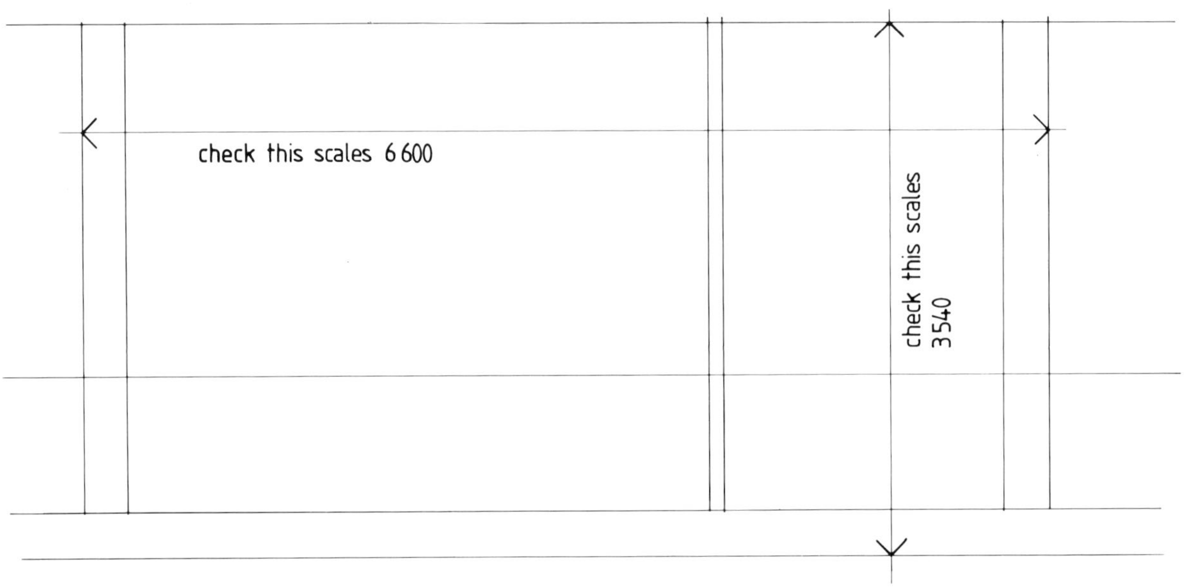

CROSS SECTION A–A	NAME	D 11/1 1st STAGE
	DATE	
	SCALE 1:50	

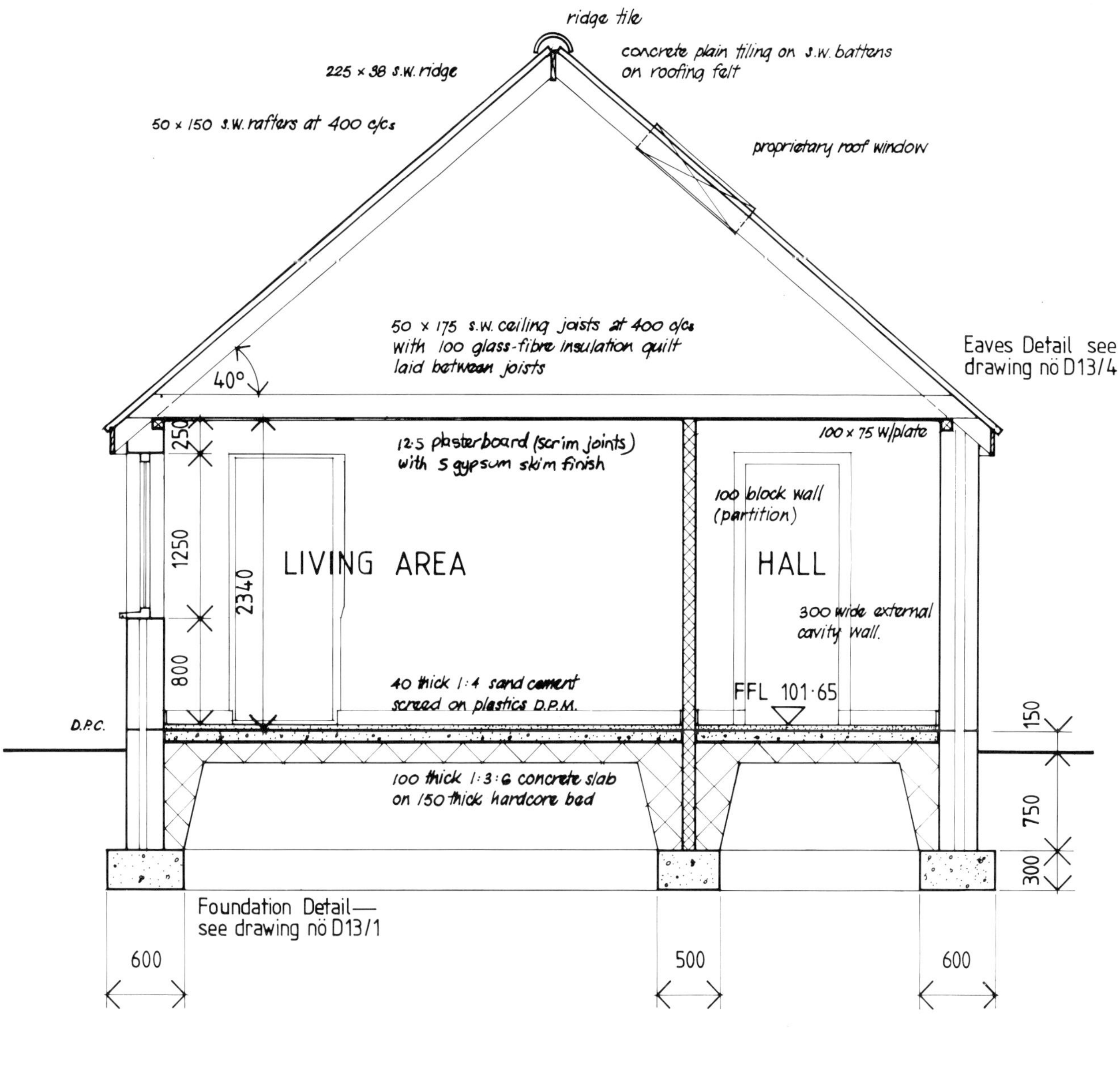

CROSS SECTION A-A

NAME

DATE

SCALE 1:50

D11/1

Draw a horizontal line denoting the top face of the ceiling joists. This line is 175 mm above the top of the ceiling.

Finish drawing the arrangement of the eaves, as shown on drawing no. D11/1.

Finish drawing the arrangement at the ridge as shown on drawing no. D11/1.

Draw a horizontal line defining the top face of the concrete floor slab. This line is 40 mm below the finished floor level (FFL). Draw a horizontal line defining the bottom face of the concrete floor slab. This line is 100 mm below the top face of the floor slab.

Draw a horizontal line defining the bottom face of the hardcore bed. This line is 150 mm below the top face of the hardcore bed.

Complete the foundations and lower parts of walls and partitions, including the hardcore, as shown on drawing numbers D11/1 and D13/1.

Draw the window on the left hand external wall as shown on drawing no. D11/1, and the door seen in elevation.

Draw the dimension lines.

Draw the guide lines for the lettering and dimensions.

Final lines
Line in the drawing starting from the top left hand corner of the sheet and working downwards and across to the bottom right hand corner. Use thick lines for the external and internal faces of all walls and partitions and all faces of the concrete floor slab and foundations. Use thin lines for all other lines including dimension lines.

Lettering
Add lettering and dimensions between guide lines. Complete the title panel.

Exercise 11.2 Practising tracing in ink

Trace in ink, on a sheet of A4 paper the drawing produced for Exercise 11.1.

Self Assessment Guide

In order to satisfy yourself that you have completed the exercises to an acceptable standard, study your drawings and check that you can answer YES to the following questions.

1 Do your drawings look the same as drawing no. D11/1 and include exactly the same information?
2 Are your drawings accurately drawn to a scale of 1:50?
3 Have thick lines been used to define the walls, floor slab and partitions, with thin lines used in all other positions?
4 Are the rafters drawn at an angle of exactly 40 degrees?

Now check with your drawings that you can also answer YES to all the questions listed in the 'Standard Self Assessment Questions' at the back of the book (page 48).

12

Location Elevations

Introduction

An elevation is a view you get if you look in a horizontal direction at the vertical side, or face, of a building or object. When drawing an elevation you need to take the horizontal dimensions from the plans and the vertical dimensions from the sections.

An example of a location elevation is shown on drawing no. D12/1.

Purpose

The purpose of a location elevation is to (a) show the external faces of the building; and (b) locate the door and window openings and other features of the building.

Scale

Location elevations are drawn to the same scales as the floor plans.

Dimensions

As a rule dimensions will not be shown on elevations of buildings. Instead this information will be provided on the plans and sections.

Depicting Materials

Horizontal lines are often used to indicate the courses of bricks and tiles, but it is not necessary to show every course.

Sub-Titles

If a drawing sheet contains more than one elevation a sub-title is required to every elevation to show which side of the building it refers to. Occasionally the terms 'front elevation', 'back elevation' and 'side elevation' are used. Generally however, particularly on large projects, elevations are described by their aspects—e.g. 'north elevation', 'south elevation' etc. Sometimes other terms such as 'access elevation' and 'garden elevation' may be used.

Lettering

Normally there will be less lettering on elevations than on plans and sections.

As well as sub-titles, some notes will be required describing the materials used on the outside of the building.

Lines

Variation of line thicknesses are often used on elevations of buildings in order to make them look as attractive as possible. Shading is often introduced to indicate shadow lines.

In this initial introduction to construction drawing it is suggested that you use thin lines throughout with the exception of the *base line*. The base line is the bottom line of an elevation representing the line of the ground.

Exercise 12.1 Elevations

Requirement
Draw the elevations shown on drawing numbers D12/1, D12/2, D12/3 and D12/4, in pencil, on a sheet of white A2 paper. The scale is to be 1:50.

Preliminaries
As this exercise is to be drawn on a sheet of A2 paper, the arrangement of the title panel and border will generally be similar to Fig. 2.2.

Construction lines
First draw the north west elevation shown on drawing no. D12/1 in the top left hand corner of your sheet of paper.

Draw the base line.

Draw the two vertical lines defining the corners of the building. These are 6600 apart.

Draw the soffit line of the roof eaves. This is 2200 above the base line.

Draw the external shape of the roof by transferring the measurements and angles from section A—A on drawing no. D11/1.

Draw the windows by transferring the measurements from the floor plan (drawing no. D10/1), section (drawing no. D11/1) and north west elevation (drawing no. D12/1).

Draw the barge board by referring to drawing no. D12/1.

Now draw the south east elevation shown on drawing no. D12/2 in the top right hand corner of your sheet of paper. The procedure to be followed is similar to that used for drawing the north west elevation.

Now draw the south west elevation shown on drawing no. D12/3 in the bottom left hand corner of your sheet of paper.

Draw the base line.

Draw the two vertical lines defining the corners of the building. These are 9100 apart.

Draw the soffit line of the roof eaves. This is 2200 above the base line.

Draw the external shape of the roof by referring to section A—A (drawing no. D11/1) and south west elevation (drawing no. D12/3).

Draw the windows and doors by transferring the measurements from the floor plan (drawing no. D10/1), section A—A (drawing no. D11/1) and south west elevation (drawing no. D12/3).

Draw the roof windows by referring to section A—A (drawing no. D11/1) and south west elevation (drawing no. D12/3).

Draw the fascia, gutter, rainwater pipe and ridge by referring to the south west elevation (drawing no. D12/3).

Now draw the north east elevation shown on drawing no. D12/4 in the bottom right hand corner of your sheet of paper. The procedure to be followed is similar to that used for drawing the south west elevation.

Draw the lines to indicate the brickwork courses and roof tile course.

Draw the guide lines for the lettering.

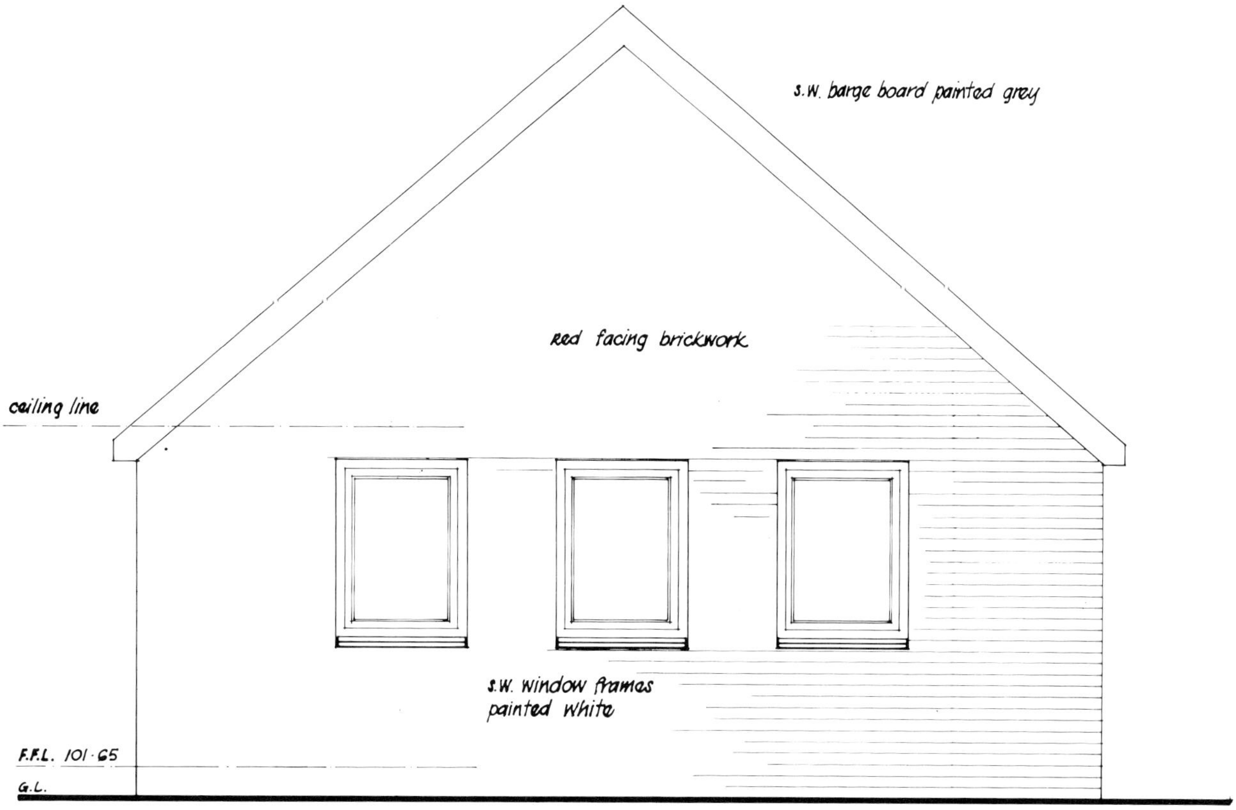

NORTH WEST ELEVATION	NAME	D12/1
	DATE	
	SCALE 1:50	

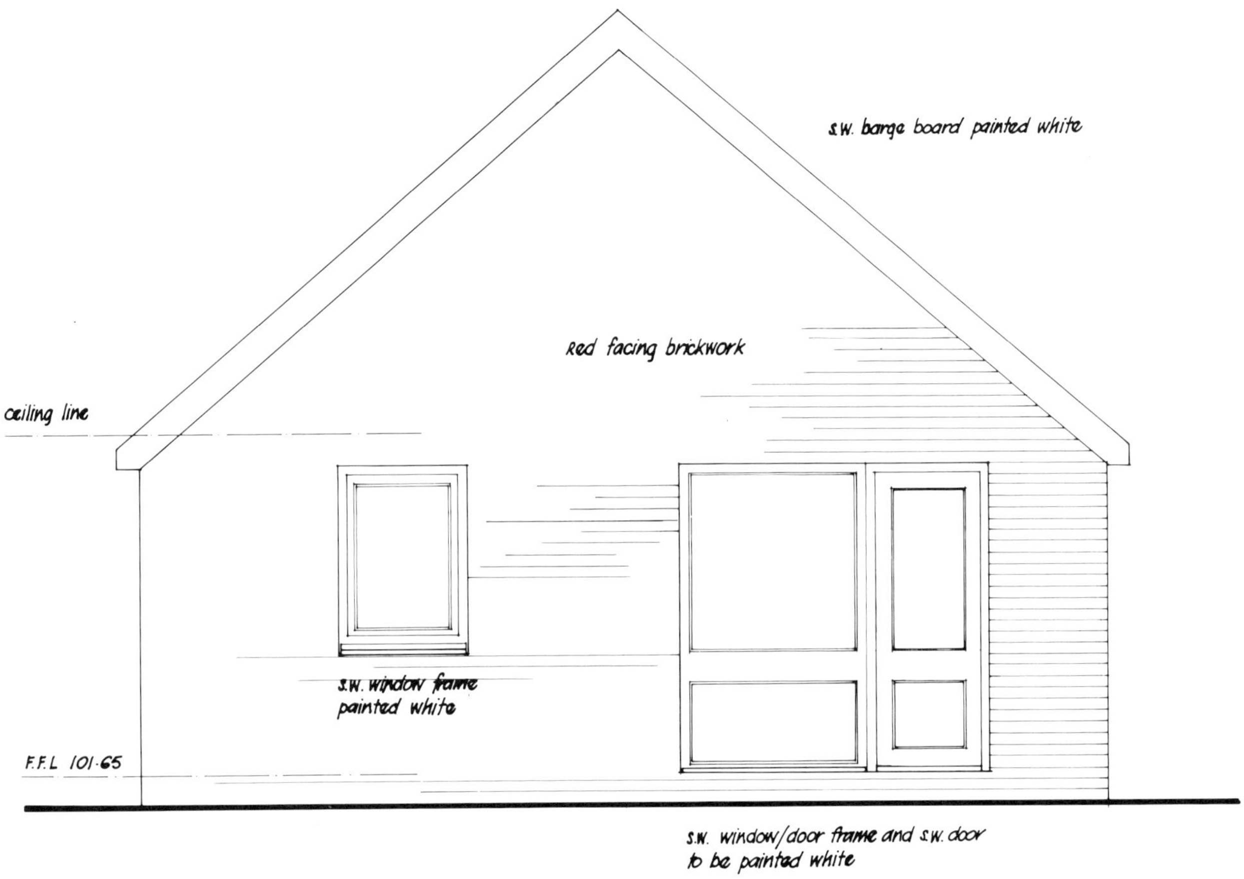

SOUTH EAST ELEVATION

NAME

DATE

SCALE 1:50

D12/2

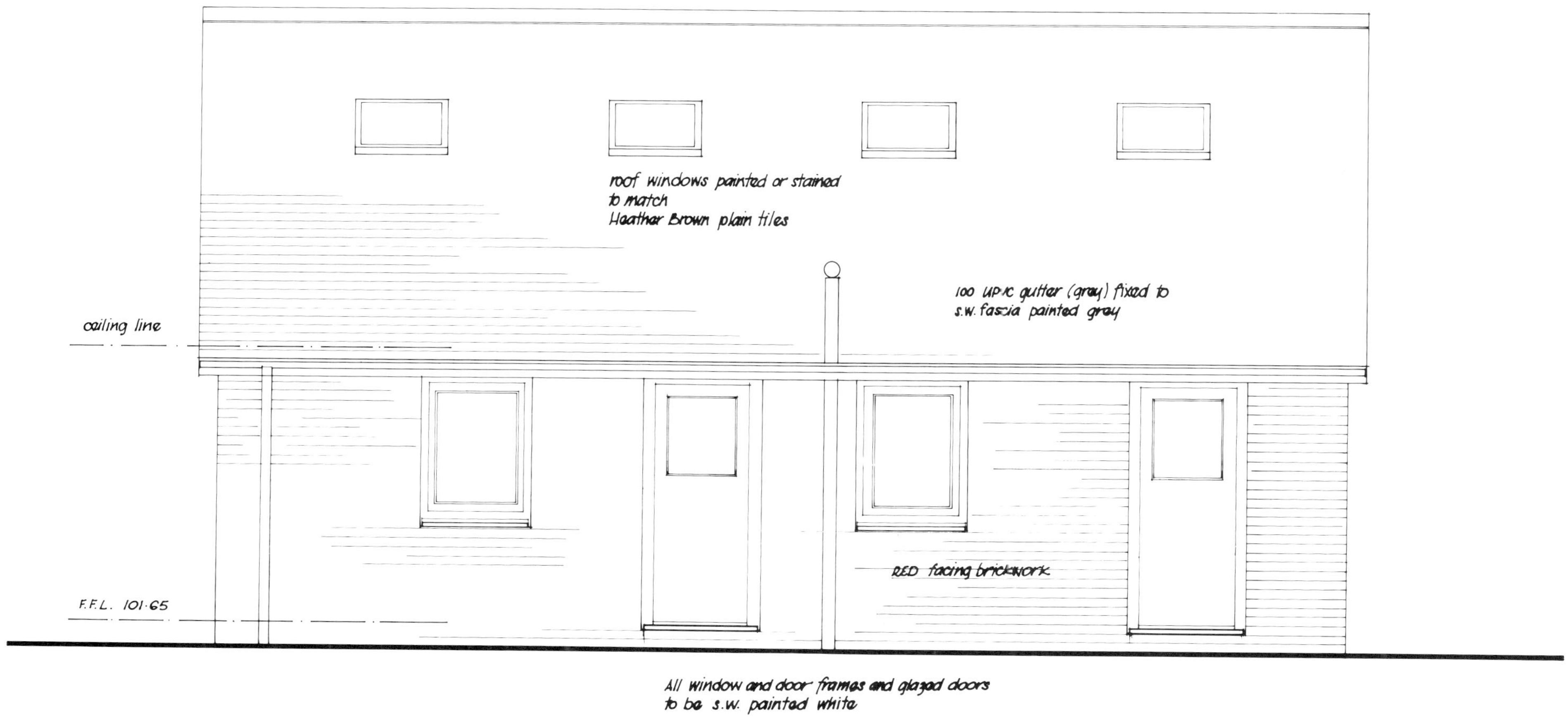

SOUTH WEST ELEVATION

NAME	D12/3
DATE	
SCALE 1:50	

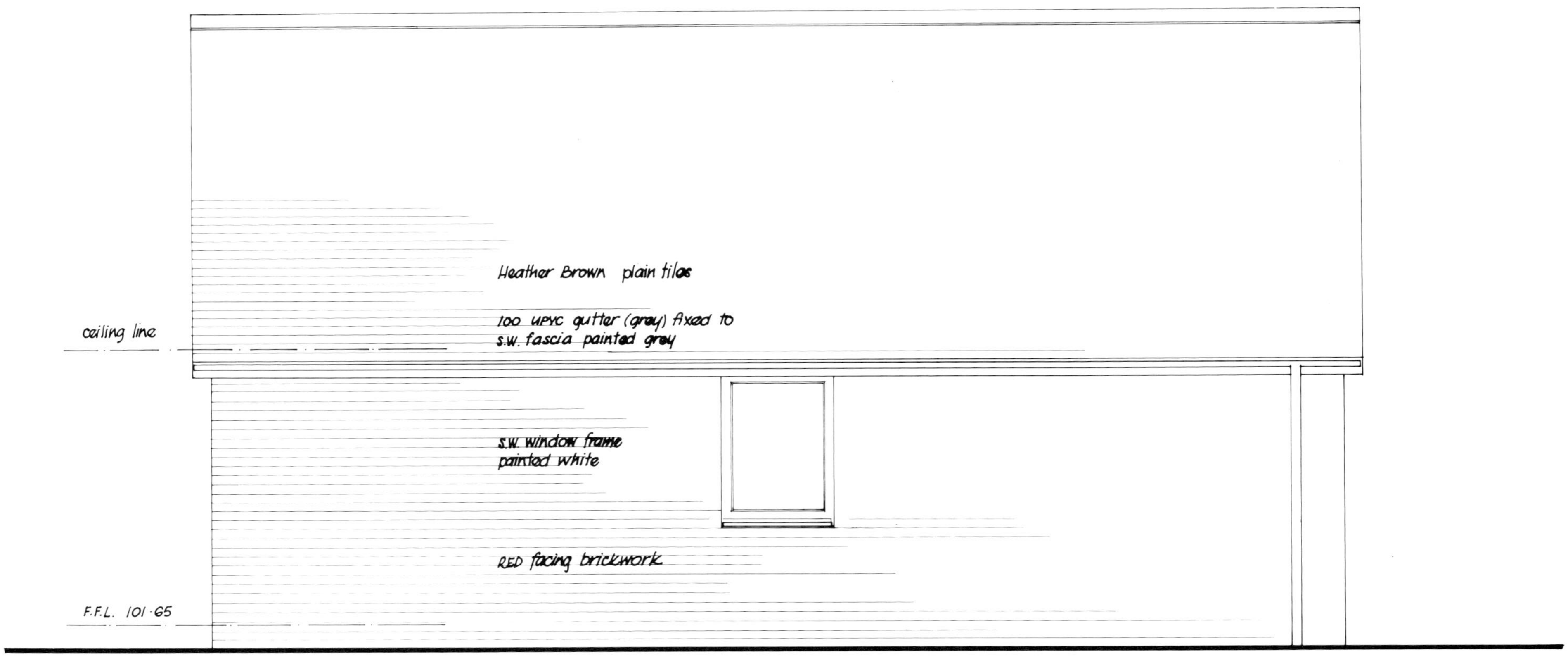

NORTH EAST ELEVATION	NAME	D12/4
	DATE	
	SCALE 1:50	

Final lines
Line in the drawing in the usual manner. Use thick lines for the base lines, and thin lines elsewhere.

Lettering
Add lettering between guide lines. Complete the title panel.

Exercise 12.2 Practising tracing in ink

Trace in ink on a sheet of A2 paper the drawing produced for Exercise 12.1.

Self Assessment Guide

In order to satisfy yourself that you have completed the exercises to an acceptable standard, study your drawings and check that you can answer YES to the following questions.

1. Do your elevations look the same as those on drawing numbers D12/1, D12/2, D12/3 and D12/4 and include exactly the same information?
2. Is your drawing accurately drawn to a scale of 1:50?
3. Is the gable roof drawn to an angle of exactly 40 degrees?
4. Have thick lines been used for the base lines, with thin lines used in all other positions?

Now check with your drawings that you can also answer YES to all the questions listed in the 'Standard Self Assessment Questions' at the back of the book (page 48).

13

Assembly Drawings

Introduction

Assembly drawings provide precise, detailed information as to the construction of buildings, including matters such as the fixing of materials, components and elements. They usually consist of sectional plans and vertical sections, but it will be appropriate on occasions to use other methods, including isometric projection and elevations.

An example of an assembly drawing is shown on drawing no. D13/1.

On some smaller projects it may not be necessary to produce assembly drawings, as the assembly information can be given on the location drawings, particularly when they are drawn to a scale of 1:50.

Purpose

The purpose of an assembly drawing is to

- show the construction of individual elements of structure such as foundations, walls, floors and roofs
- show the arrangement where two elements meet each other—e.g. the junction between a wall and a roof, and between a column and a wall
- provide a reference as to where more detailed information about a particular part of the construction is provided.

Scale

A wide variety of scales may be used for assembly drawings. Common ones are 1:2, 1:5, 1:10 and 1:20.

Dimensions

It is important to dimension assembly drawings fully, although it is not generally necessary to repeat dimensions already given on location floor plans, sections and elevations.

As with other drawings, dimensioning methods should be standardized. Figures should be placed above and in the same direction as the dimension lines, so that they can be read from the bottom and right hand side of the drawing.

Lines

Thick lines should be used to define key elements—for example the inside and outside faces of walls, the faces of structural concrete etc.

Thin lines should be used to indicate parts of the construction which are seen in elevation—e.g. roof rafters and ceiling joists in a cross section. (see D13/4).

Non structural items—for instance a plasterboard ceiling—should be indicated by thin lines whether seen in elevation or section.

Hatching

Hatching is generally used on assembly drawings to indicate the composition of materials which are shown in section.

Annotation

Sufficient notes should be provided to clarify anything which would not otherwise be obvious. However there is no need to duplicate information given elsewhere. Indeed it can be counter-productive, as undue repetition may result in drawings becoming crowded, confusing and difficult to read.

Due to its importance, it is worth repeating that all lettering must be neat, easy to read, and preferably attractive to look at. Always use guide lines, but make them very faint—barely visible.

Reference Points

It is important to include relevant grid lines and other references on assembly drawings, so they can be related to other drawings, particularly location drawings.

Examples of Assembly Drawings

Assembly drawings provide information to contractors which enable them to construct buildings on site. They include the assembly of both structural and non structural elements, components and materials. Some examples are given below.

Drawing no. D13/1 is an assembly drawing for an external wall foundation. It shows the relationship between the foundation, wall and ground floor.

Drawing no. D13/2 is an assembly drawing of an internal door opening, and is in the form of a vertical section. It shows the partition, lintel, door lining, door and wall finish.

Drawing no. D13/3 is an assembly drawing of an external window opening, and is in the form of a sectional plan. It shows the wall, window and wall finishes, and includes information about the damp proof course.

Drawing no. D13/4 is an assembly drawing of a roof eaves, and is in the form of a vertical section. It shows the roof structure and covering, and various finishing items.

Some further examples of assembly drawings are listed below.

- Sectional plan showing the exact relationship between a wall and a column.
- Vertical section at the junction between a floor and a wall.
- Vertical section between a wall and a flat roof, including details at the roof edge.
- Vertical sections at window and door heads and cills.
- An isometric view of a dormer window in a roof.
- An elevation showing the setting out for an arched opening.

Exercise 13.1 Foundation section

Requirement
Copy drawing no. D13/1 on a sheet of A4 paper. The scale is to be 1:10.

It is left to your discretion in this and future exercises, as to whether you use pencil or ink, apart from tracing exercises which should be in ink.

Preliminaries
Undertake the preliminaries as described for previous exercises.

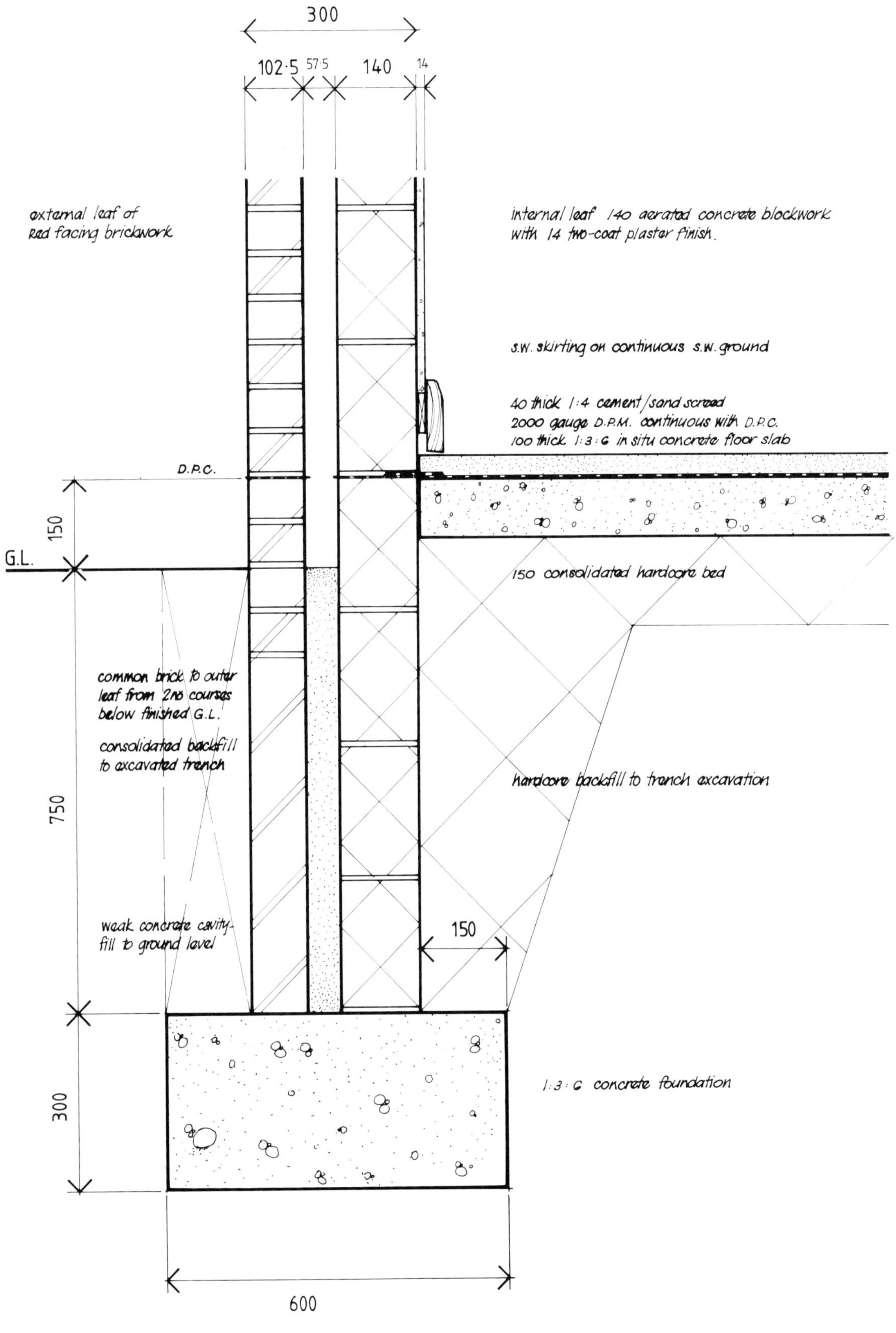

EXTERNAL WALL – FOUNDATION DETAIL	NAME	D 13/1
	DATE	
	SCALE 1:10	

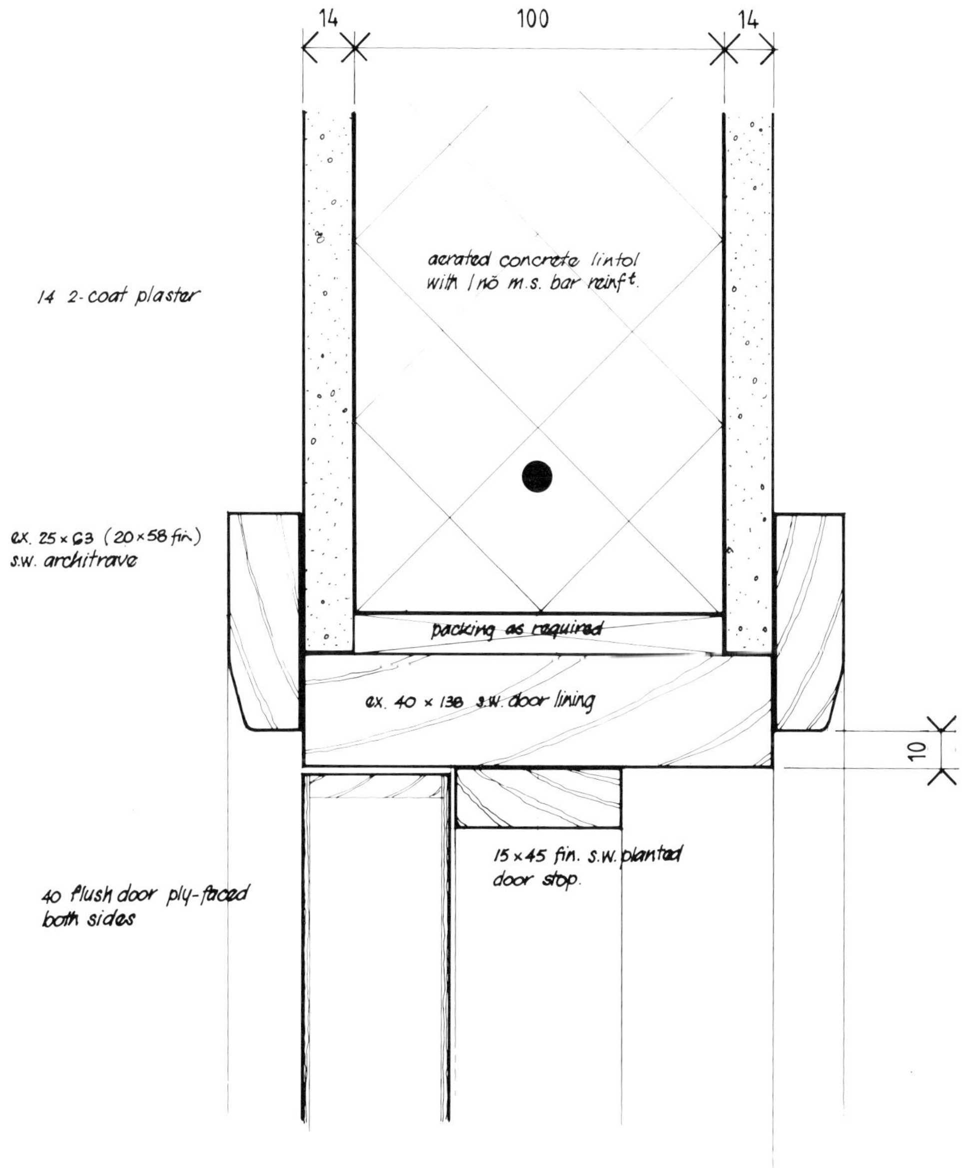

INTERNAL DOOR OPENING HEAD DETAIL	NAME	D 13/2
	DATE	
	SCALE 1:2	

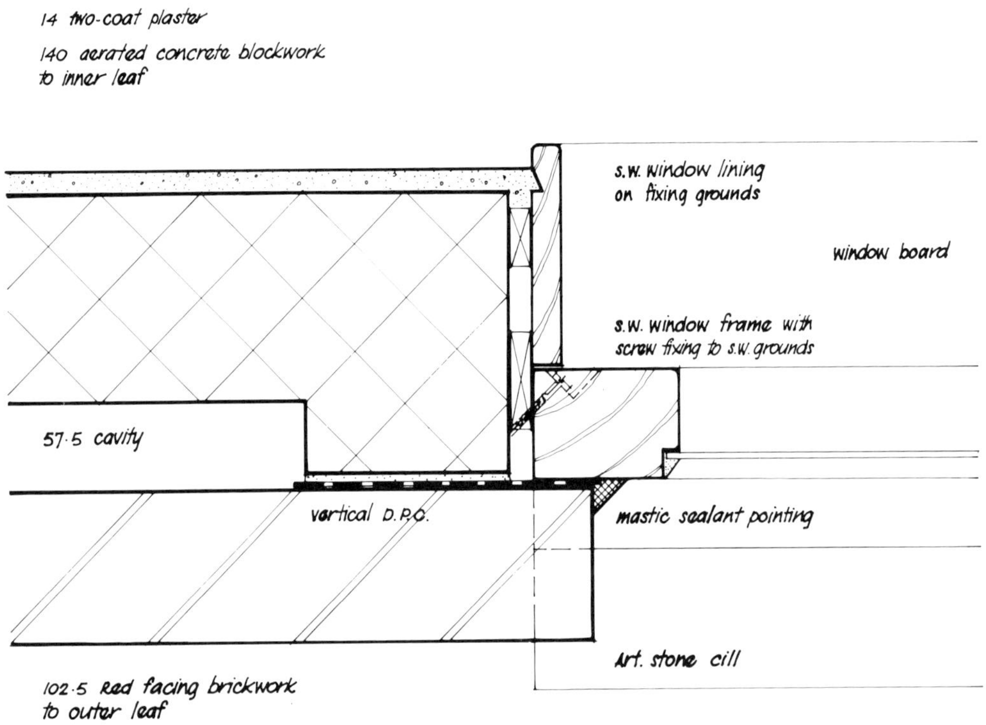

WINDOW - JAMB DETAIL	NAME	D 13/3
	DATE	
	SCALE 1:5	

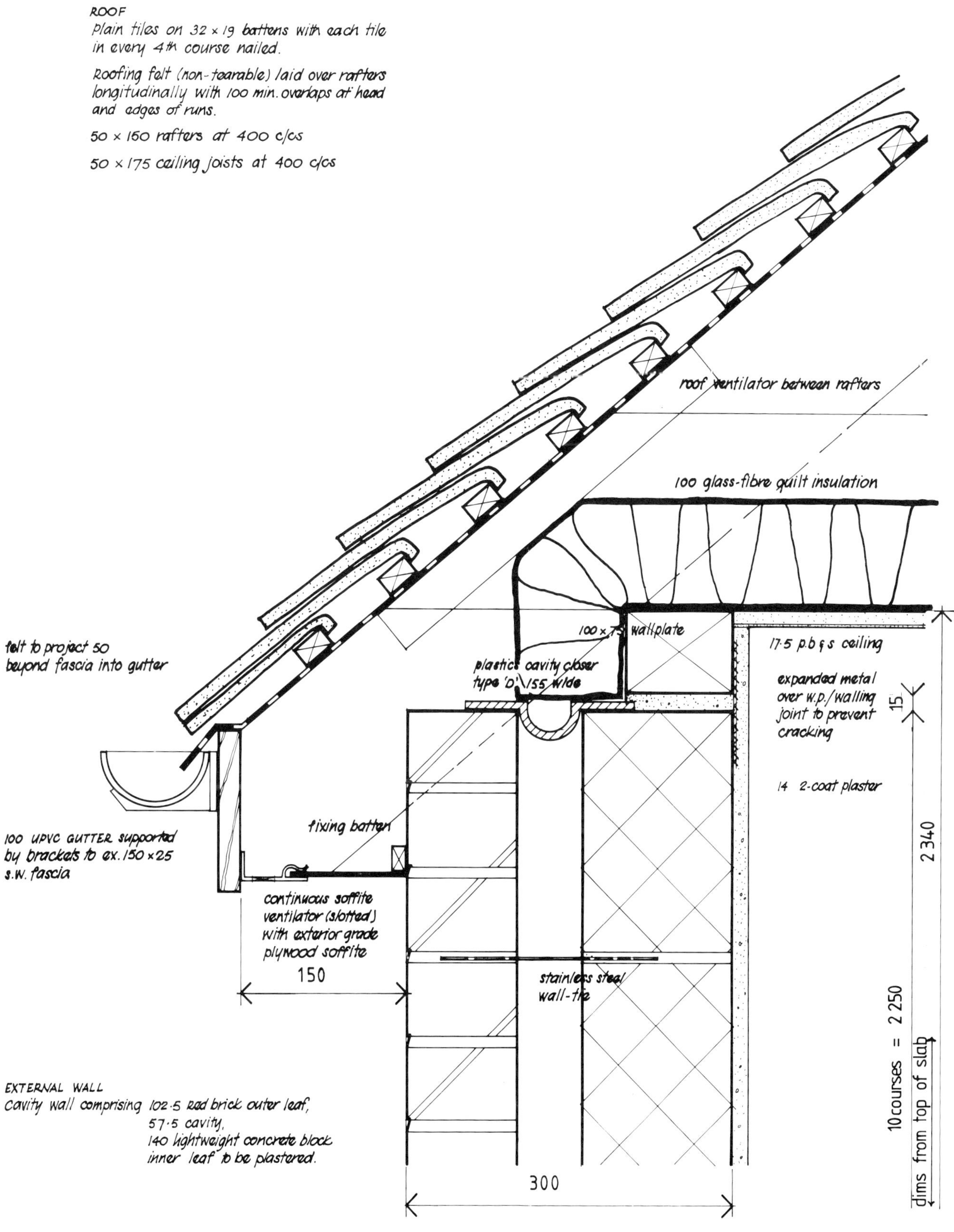

EAVES DETAIL

NAME

DATE

SCALE 1:5

D 13/4

Construction lines
Draw the four vertical lines representing the faces of the brickwork and blockwork to the wall.

Draw two horizontal lines representing the top and bottom faces of the foundation.

Measure offsets of 150 along the top horizontal line, and draw two vertical lines representing the sides of the foundations.

Draw the remaining horizontal lines, followed by the remaining vertical and slanting lines.

Draw the dimension lines and guide lines for lettering.

Final lines
Line in the drawing in the usual manner. Use thick lines for the foundations, faces of floor slab and faces of wall. Use thin lines elsewhere.

Lettering
Add lettering and dimensions between guide lines. Complete the title panel.

Exercise 13.2 Door opening section

Requirement
Copy drawing no. D13/2 on a sheet of A4 paper. The scale is to be 1:2.

Preliminaries
Undertake the preliminaries as described for previous exercises.

Construction lines
Draw the horizontal line and the two vertical lines representing the faces of the concrete lintel.

Draw the two vertical lines representing the outside faces of the plaster.

Draw the remaining horizontal lines, followed by the remaining vertical lines, and slanting lines.

Draw dimension lines and guide lines for lettering.

Completion
Complete the drawing by lining in and lettering, as described for previous exercises. Thin lines can be used throughout, or alternatively the lines defining the timber faces can be thickened.

Exercise 13.3 Window jamb plan

Requirement
Copy drawing no. D13/3 on a sheet of A4 paper. The scale is to be 1:5.

Preliminaries
Undertake the preliminaries as described for previous exercises.

Construction lines
Draw the four horizontal lines and the two vertical lines representing the faces of the brickwork and blockwork to the wall.

Draw the horizontal and vertical lines representing the window on section and plan.

Draw the remaining horizontal lines, followed by the remaining vertical lines, and slanting lines.

Draw guide lines for lettering.

Completion
Complete the drawing by lining in and lettering, as described for previous exercises. It is suggested that thick lines be used for exposed wall faces, and thin lines elsewhere.

Exercise 13.4 Eaves section

Requirement
Copy drawing no. D13/4 on a sheet of A4 paper. The scale is to be 1:5.

Preliminaries
Undertake the preliminaries as described for previous exercises.

Construction lines
It is suggested that the various parts of the detail be drawn in the following order.

Main faces of the brick/block wall.
Wall plate. Rafters. Ceiling joists.
Fascia and soffit, including cradling.
Gutter, including brackets.
Wall and ceiling finishes. Insulation.
Detailed work to cavity wall.
Roofing felt, battens and tiles.
Draw guide lines for lettering.

Completion
Complete the drawing by lining in and lettering, as described for previous exercises. It is suggested that thick lines be used for exposed wall faces, and thin lines elsewhere.

Exercise 13.5 Practising tracing in ink

Trace in ink on a sheet of A4 paper the drawing produced for Exercise 13.4.

Self Assessment Guide

In order to satisfy yourself that you have completed the exercises to an acceptable standard, study your drawings and check that you can answer YES to the following questions.

Exercise 13.1

1 Does your drawing look exactly the same as drawing no. D13/1 in this book?
2 Is your drawing accurately drawn to a scale of 1 : 10?
3 Have thick lines been used to define the faces of the concrete floor slab, foundations and walls, with thin lines used in all other positions?

Exercise 13.2

1 Does your drawing look exactly the same as drawing no. D13/2 in this book?
2 Is your drawing accurately drawn to a scale of 1 :2?

Exercise 13.3

1 Does your drawing look exactly the same as drawing no. D13/3 in this book?
2 Is your drawing accurately drawn to a scale of 1 :5?

Exercise 13.4

1 Does your drawing look exactly the same as drawing no. D13/4 in this book?
2 Is your drawing accurately drawn to a scale of 1 :5?

Now check with each of your drawings that you can also answer YES to all the questions listed in the 'Standard Self Assessment Questions' at the back of the book (page 48).

14

Component Drawings

Introduction

A component drawing provides detailed information about the nature and manufacture of a specific item incorporated in a building. This is in contrast to an assembly drawing which shows several parts, or a location drawing which provides general information. Components include things such as skirtings and lintels, as well as larger items manufactured off-site, such as windows and kitchen cupboards.

An example of a component drawing is shown on drawing no. D14/1.

On some smaller projects it may not be necessary to produce component drawings, as component information can be given on assembly drawings. They should only be provided where it is not feasible to provide the information in some other way.

Purpose

The purpose of a component drawing is to (a) show the nature, shape, assembly method and further details of components, required by the manufacturer and others; and (b) provide additional information which cannot be conveniently given on location or assembly drawings.

Scale

A wide variety of scales may be used for component drawings. Common ones are 1:1, 1:2, 1:5, 1:10, 1:20 and 1:50.

Dimensions

Because component drawings are used by contractors, subcontractors and suppliers to manufacture items for incorporating in a building, it is essential that they should be accurately drawn to scale and be dimensioned in a precise and detailed way.

Lines

Thick lines should generally be used to define faces of items viewed in section, such as the cill section on drawing no. D14/2.

Thin lines should be used where elevational views are included, such as the door elevation on drawing no D14/1.

Hatching

Hatching is commonly used on component drawings to indicate the composition of materials which are shown in sections.

Annotation

Component drawings should be fully annotated, in such a way as to provide clear, precise information to the manufacturer and others.

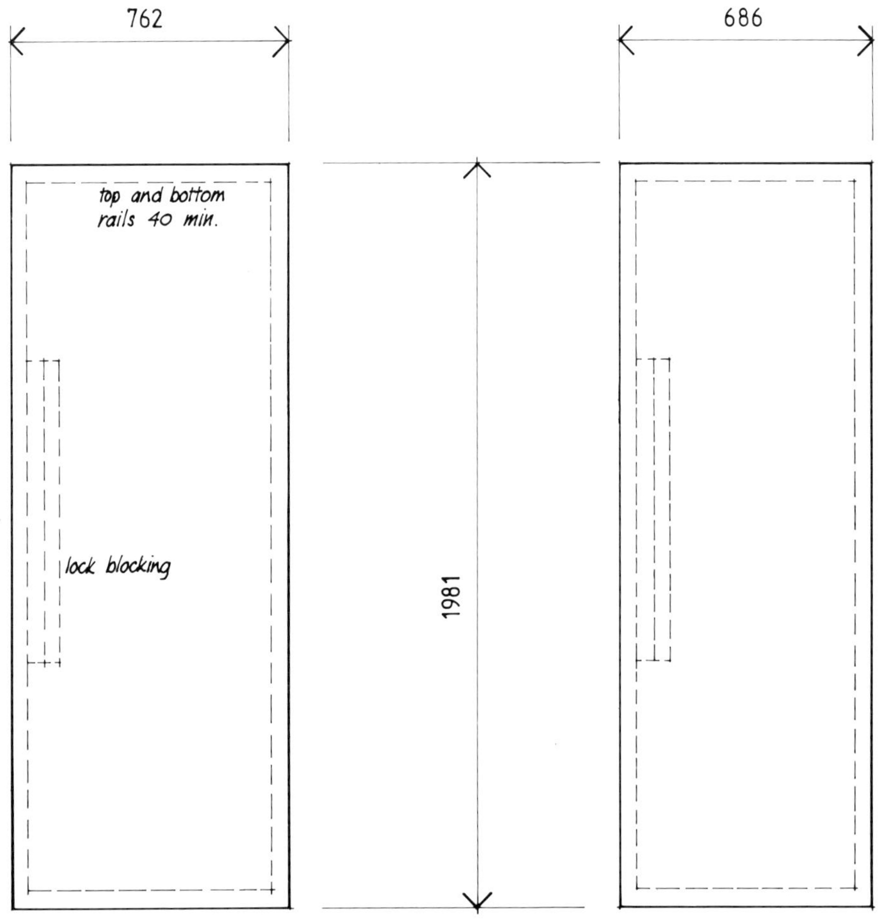

Doors D4, D5 and D8

Doors D6 and D7

ELEVATIONS

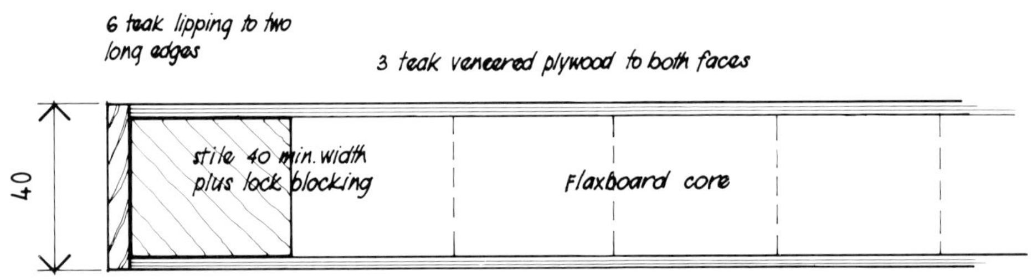

PLAN AT DOOR EDGE

INTERNAL DOORS	NAME	D14/1
	DATE	
	SCALE 1:20 and 1:2	

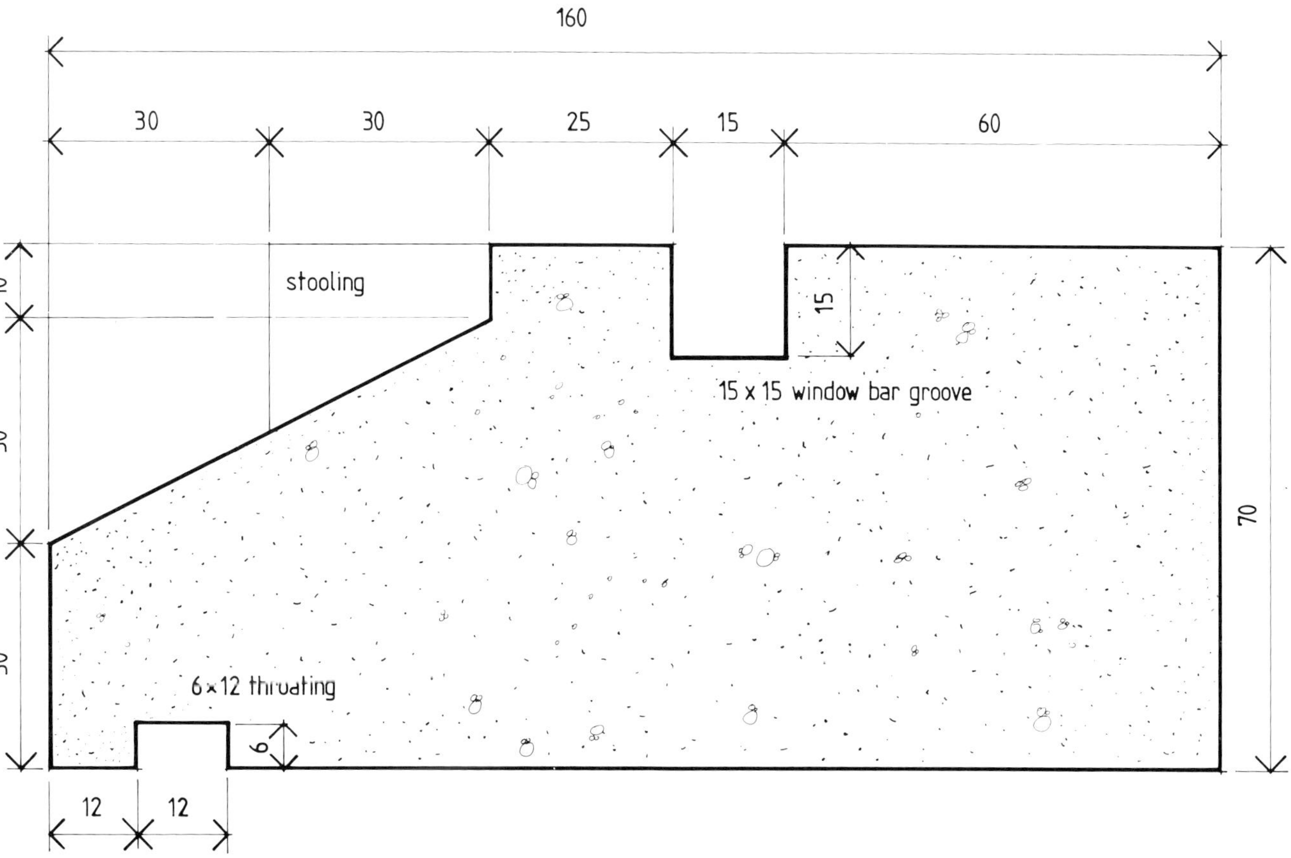

SECTION

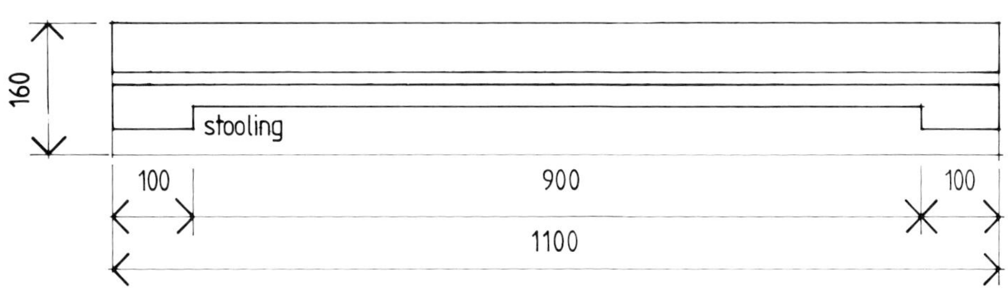

PLAN

PRECAST CONCRETE CILL	NAME	D14/2
	DATE	
	SCALE 1:1 and 1:10	

Examples of Component Drawings

Some examples of component drawings are given below.

Drawing no. D14/1 is a component drawing for doors, and is in the form of elevations and a part sectional plan.

Drawing no. D14/2 is a component drawing of a concrete cill, and is in the form of a plan view and a vertical section.

Further examples of component drawings are drawings produced to provide information about windows, cladding panels, partition units, ceiling panels, kitchen units, cupboard fittings, roof lights and softwood trim.

Exercise 14.1

Doors

Requirement
Copy drawing no. D14/1. Scales are to be 1:2 and 1:20.

Preliminaries
Undertake the preliminaries as described for previous exercises.

Construction lines
It is suggested that the various parts be drawn in the following order.

Main outlines of the two door types, indicated by full lines on the elevations.

Detail on the elevations, indicated by broken lines.

Door outline on plan.

Detail of plan.

Hatching.

Dimension lines.

Guide lines for lettering.

Completion
Complete the drawing by lining in, and lettering, as described for previous exercises. It is suggested that thin lines be used throughout.

Exercise 14.2

Concrete cill

Requirement
Copy drawing no. D14/2. Scales are to be 1:1 and 1:10.

Preliminaries
Undertake the preliminaries as described for previous exercises.

Construction lines
It is suggested that the various parts be drawn in the following order.

Outline of the cill in section, commencing with the bottom line.

Stooling of the cill, shown as an elevational view on the section.

Outline of the cill on plan.

Remaining lines on plan.

Dimension lines.

Guide lines for lettering.

Completion

Complete as for the previous exercise. Use thick lines for the cill profile seen in section.

Exercise 14.3 **Practising tracing in ink**

Trace in ink on a sheet of A4 paper the drawing produced for Exercise 14.2.

Self Assessment Guide

In order to satisfy yourself that you have completed the exercises to an acceptable standard, study your drawings and check that you can answer YES to the following questions.

Exercise 14.1 Internal doors

1 Does your drawing look exactly the same as drawing no. D14/1 in the book?

2 Are your elevations accurately drawn to a scale of 1:20 and the plan detail accurately drawn to a scale of 1:2?

Exercise 14.2 Precast concrete cill

1 Does your drawing look exactly the same as drawing no. D14/2 in the book?

2 Is the plan accurately drawn to a scale of 1:10 and the section accurately drawn to a scale of 1:1?

Now check with each of your drawings that you can also answer YES to all the questions listed in the 'Standard Self Assessment Questions' at the back of the book (page 48).

15

Structural Engineering Drawings

Generally

Introduction

Structural engineering drawings, like all technical drawings, are a means of communication, and as such are intended to provide specific information to particular people. They help those whose job it is to fabricate, erect, supervise and integrate the structural engineering work. The different needs and priorities of these people influence the method of providing the information, which will sometimes vary from architectural drawing techniques previously discussed. Some of the implications are mentioned below.

Types of Structural Engineering Work

Structural engineers and structural engineering technicians produce drawings for a wide range of structural methods, including structural steelwork, reinforced concrete using *in situ*, precast and prestressed concrete, structural brickwork and blockwork, and structural timber work.

In this introduction to construction drawing consideration will only be given to a few simple examples of framed structures in structural steelwork and *in situ* reinforced concrete.

Structural Detailing

The process of preparing working drawings for structural engineering work is generally referred to as structural detailing. The general principle followed is to break down the total structure into individual elements such as columns, beams, floor slabs etc., and then to detail each element in turn.

Standardization

Chapter 5 dealt with the importance of employing a common graphic language in the preparation of construction drawings, and following agreed standards in respect of lines, hatching, symbols etc. This is particularly important in the case of structural engineering drawings.

Structural engineering generally uses standard steel sections in structural steelwork, and standard reinforcement in reinforced concrete work; as well as standard techniques with regard to items such as hole sizes and spacings, and reinforcement cover. The various rules are contained in the British Standards and other documents, and draughtspersons must become familiar with them as their careers develop.

Structural Steelwork

Types of Drawings

The three main groups of drawings for illustrating steelwork structures are general arrangement drawings, fixing details, and details of individual members.

General arrangement drawings include steel framing plans, elevations and sections. Drawing no. D15/1 is an example of a steelwork floor framing plan. Steel framing plans indicate the positions and sizes of beams at a specific floor or roof level, together with the positions and sizes of columns. Elevations and sections will show columns and beams as well as additional members such as wind bracing.

Fixing details provide information on the fixing of members to each other, or to different parts of the structure. Examples are the fixing of a column to a foundation, a connection between a beam and a column, the connection of one beam to another beam, and the splicing of similar members to each other. Drawing no. D15/2 is an example of a fixing detail showing the connection between a steel column and a concrete base.

Details of individual members are primarily used by the workshop to fabricate the members ready for dispatching to the site, and are often referred to as shop details. They provide information required for cutting members to the correct lengths, drilling holes, attaching plates and cleats, and forming notches. Drawing no. D15/3 is an example of a beam detail.

Scales

General arrangement framing plans, elevations and sections are generally drawn to a scale of 1:100, but 1:200 is also used occasionally.

Fixing details are drawn to a variety of scales, including 1:5, 1:10 and 1:20.

Details of individual members are also drawn to various scales, such as 1:5, 1:10, and 1:20.

Mark References for Beams and Columns

It is necessary to identify each steel member by a distinctive mark reference whenever it appears on a drawing. The same mark must also be painted on the actual member before it leaves the workshop for dispatch to the site.

A common system is to mark the horizontal grid lines on the steel framing plans with numbers—1, 2, 3 etc.—and the vertical grid lines with letters—A, B, C etc. This is illustrated on drawing no. D15/1 On the sections the floors are given a letter beginning with the lowest floor—A for ground floor, B for first floor etc. This is illustrated on Fig. 15.1.

This system enables each member to be identified. Columns are given a mark related to the grid intersections on the plans. Thus the top left hand column on drawing no. D15/1 is given the mark of A1 because it is located where grid lines A and 1 intersect. If the floor column is not continuous from the lowest floor to the roof, the reference letter of the floor where the column starts is added to the column mark. For example in the case of drawing no. D15/1, if the column ran from the second floor to the roof, the letter C would be added to the column mark, making the top left hand column A1-C. The marks for the beams are a combination of the floor reference and the grid line letter and number. For example on drawing no. D15/1 the horizontal beam in the top left hand corner is marked as C-1A. The letter C indicates it is a second floor beam; the figure 1 denotes that the beam is located at grid line 1; and the letter A denotes that it begins at grid line A.

An alternative method is to number the beams in sequence, generally beginning at the top left hand corner and working downwards towards the bottom right hand corner. The floor reference letters—i.e. A, B, C etc.—are added to the number. Under this system the beam in the top left hand corner of Fig. 15.1 is marked as C1; C denoting it is a second floor beam, and 1 indicating it was the first beam to be numbered.

Fixing Methods

The various steel members—universal beams, universal columns, rolled joists, rolled channels, tees and angles—are fixed together by welding or bolting, either in the workshop (shop connections) or on the construction site (site connections).

Standard symbols are used to indicate the type of bolts used and whether they are fixed in the shop or on site. Some common symbols are shown on Fig. 15.2, but to obtain the complete information it will be necessary to refer to the relevant British Standards.

Lines

Beams and columns on framing plans are generally indicated by single thick lines.

Columns and beams on elevations are commonly indicated by two thin lines or a single thick line.

Wind bracing is commonly indicated by a thick broken line.

Centre lines, grid lines and dimension lines use thin lines.

Dimensions

The information on dimensioning given in Chapter 4 generally applies to structural engineering drawings. Dimension lines are commonly terminated by solid arrowheads.

Sometimes running dimensions are used to position holes in steel members, as opposed to the more common practice of providing a series of independent dimensions. Both these methods are illustrated on Fig. 15.2.

***In situ* Reinforced Concrete Structures**

Types of Drawings

The three main groups of drawings for illustrating *in situ* reinforced concrete structures are general arrangement drawings, reinforcement drawings and profile drawings.

General arrangement drawings are floor plans, roof plans, sections and elevations, drawn to a small scale and providing an overall view of the work. They supply the setting out dimensions, the positions and sometimes the sizes of all the members. Drawing no. D15/4 is an example of an RC general arrangement drawing, and shows the layout of beams and columns at first floor level. A reference grid is provided similar to that previously described for a structural steel building.

Reinforcement drawings are drawn to a larger scale and give detailed information about the reinforcement, including its location in the various elements such as columns and beams. They are the means by which the steel reinforcement fixer on the construction site does what the structural engineer in the office intended when he designed the structure. There is no excuse for ambiguity, and it is essential that all drawings are easy to read, and cannot be misunderstood. A simple example of a reinforcement drawing is that at the base of a reinforced concrete column shown on drawing no. D15/5. It should be noted that the reinforcement is indicated by full thick lines on the plan and elevation. This is the normal procedure on reinforced concrete drawings.

Profile drawings will generally be required only on larger projects and provide detailed information about the various members. Their main purpose is to assist the formwork carpenter.

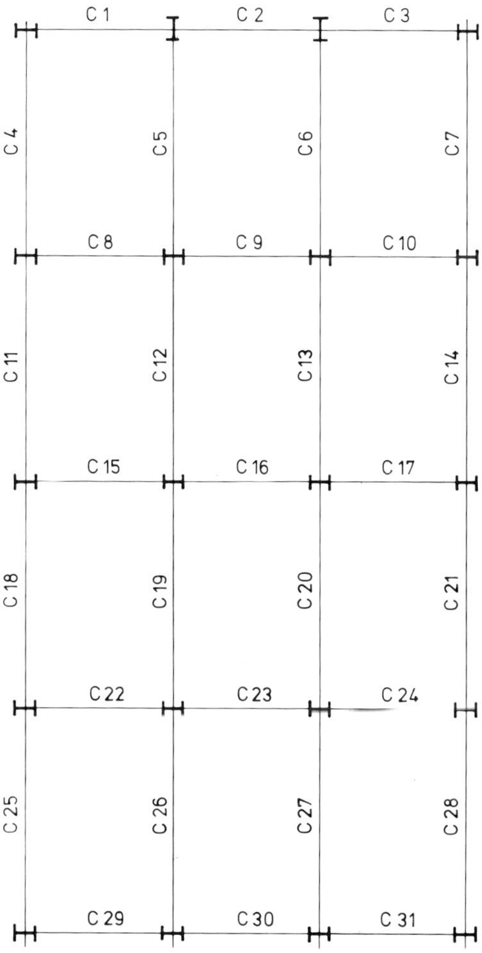

SEQUENTIAL NUMBERING OF BEAMS
SECOND FLOOR PLAN

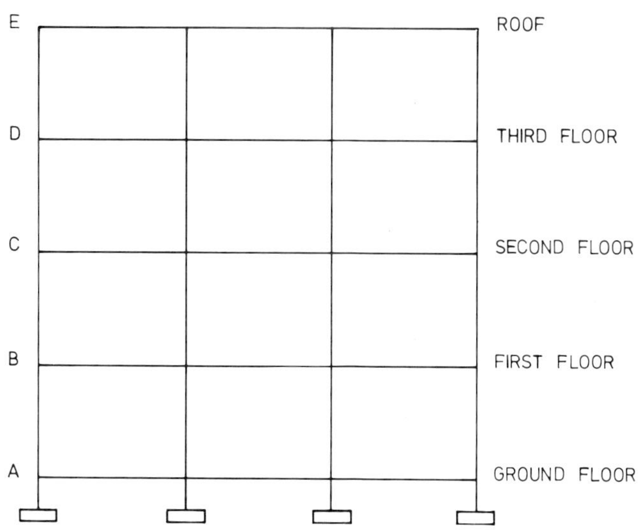

MARKING SYSTEM FOR FLOORS
SECTIONS / ELEVATIONS

Fig. 15.1

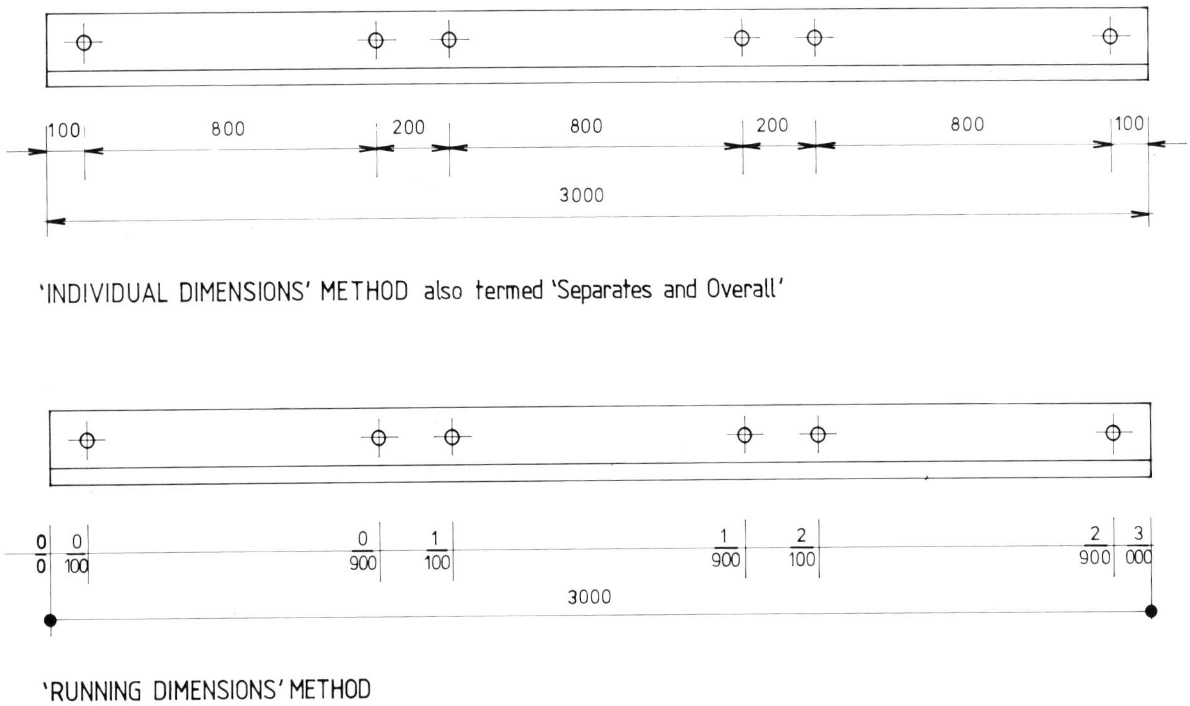

(A) DIMENSIONING OF HOLES IN STEEL MEMBER

(B) STANDARD SYMBOLS FOR BOLTS

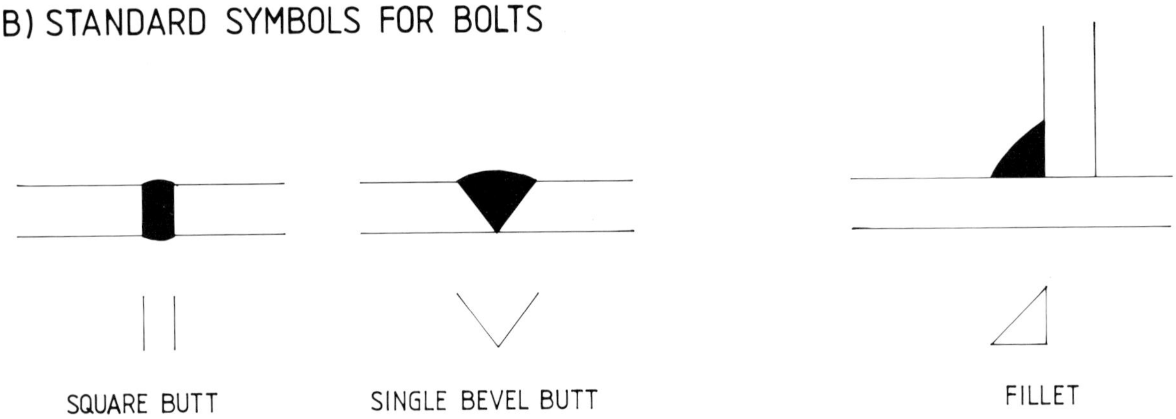

(C) TYPES OF WELDS AND STANDARD SYMBOLS

Fig. 15.2 Steelwork dimensioning and symbols

Scales

General arrangement drawings are drawn to a scale of 1:100 or 1:50.

Simple wall and slab details are drawn to a scale of 1:50. More complicated details, including those for columns and beams are generally drawn to a scale of 1:20, but sometimes a 1:10 scale is used.

Mark References for Columns and Beams

Columns are given reference marks obtained from the grid lines. On drawing no. D15/4 the column A1 is at the intersection of grid line A and grid line 1. The beams are given a notation obtained from the grid line they are on and the nearest grid line in the other direction—1a, 2b etc.

Drawing Reinforcement

In order to help the student understand a typical arrangement of reinforcement in an R.C. column base, on drawing D15/5 every reinforcement bar has been shown. However, in practice this amount of detail will not normally be provided. Generally on the plans and elevations only one bar of each type will be shown in full. For example on the plan of the column base only one bar running in a horizontal direction and one bar running in a vertical direction would be indicated; where elements are shown in section—such as the column shown in sectional plan on drawing no. D15/5—the reinforcement would be shown in full.

Reinforcement Identification

Standard abbreviations are used to provide information about the reinforcement. The following are used on drawing no. D15/5.

R mild steel round bars

T high tensile bars

B reinforcement near the bottom face of the concrete.

Each reinforcing bar on a drawing is given a notation consisting of standard abbreviations, dimensions in mm and mark numbers—i.e. 01, 02, 03 etc. This information is provided in the following sequence: number, type, size, mark, centres and location.

The meaning of the notations given to the reinforcement on drawing no. D15/5 is as follows.

9T 20 01 200B This means that there are 9 bars, which are of high tensile steel, with a diameter of 20 mm, and an identification mark of 01. The bars are spaced 200 mm apart and placed near the bottom face of the concrete base.

4R 25 03 This means that there are 4 bars, which are mild steel round bars, with a diameter of 25 mm and an identification mark of 03.

5R 6 04 200 This means that there are 5 bars, which are mild steel round bars, with a diameter of 6 mm, an identification mark of 04 and the bars are spaced 200 mm apart.

There is no need to repeat the full notation each time a bar is shown. For example on drawing no. D15/5 the full notation for the reinforcement bars in the base are given on the plan view, but not provided in full on the elevation.

Lines

The reinforcement is the most important item on reinforced concrete drawings and should therefore be drawn using the thickest lines. They are often made 0.70 mm thick, but 0.50 mm can be used if only two pen or pencil thicknesses are available. It is normal practice to draw section outlines with thicker lines than those used for plans and elevations, so 0.50 mm can be used for this purpose. Centre lines, grid lines and dimension lines are normally drawn with the thinnest lines.

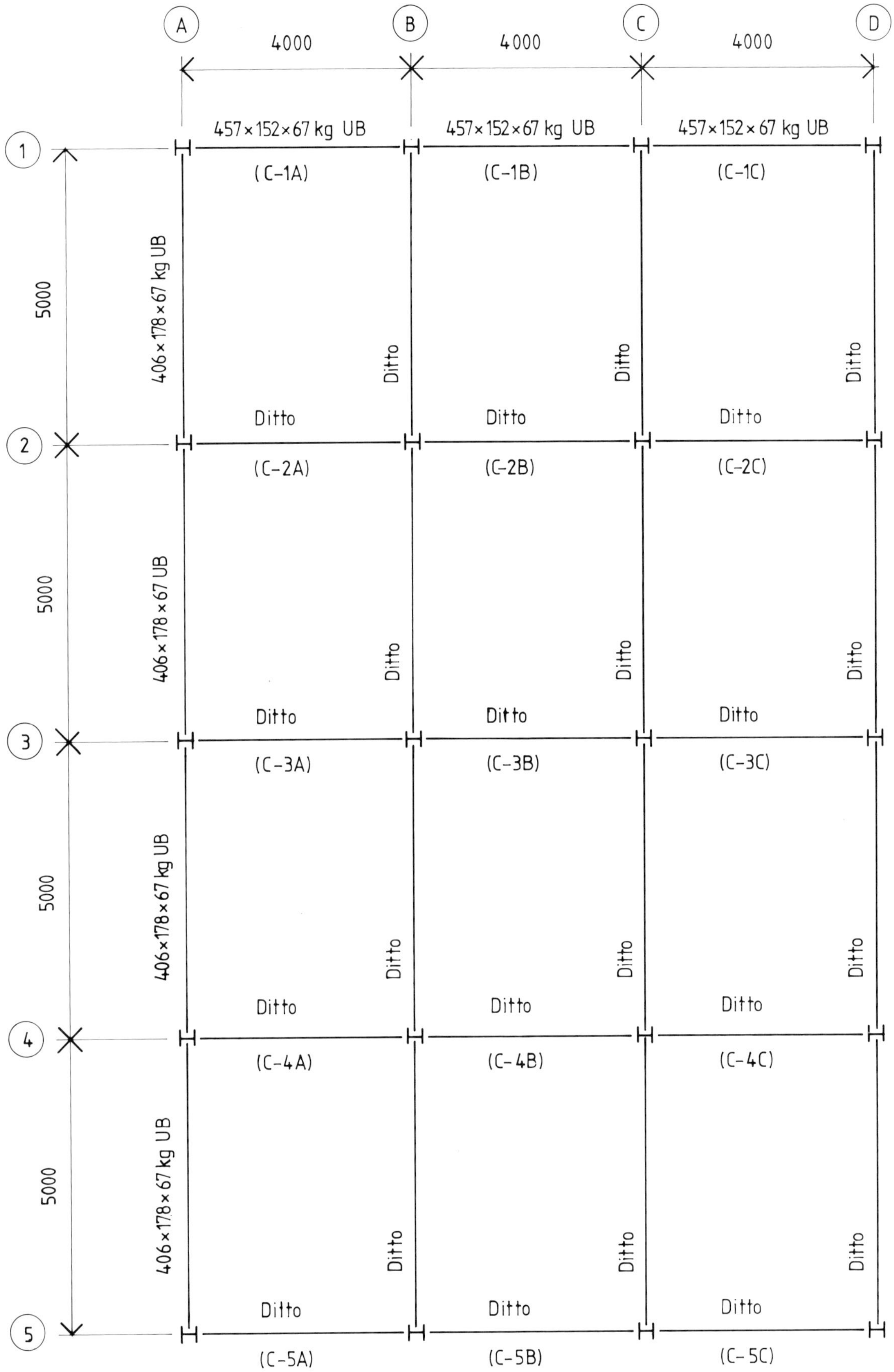

ALL COLUMNS ARE 254 × 254 × 73 kg UC

2nd FLOOR STEEL FRAMING PLAN	NAME	D15/1
	DATE	
	SCALE 1:100	

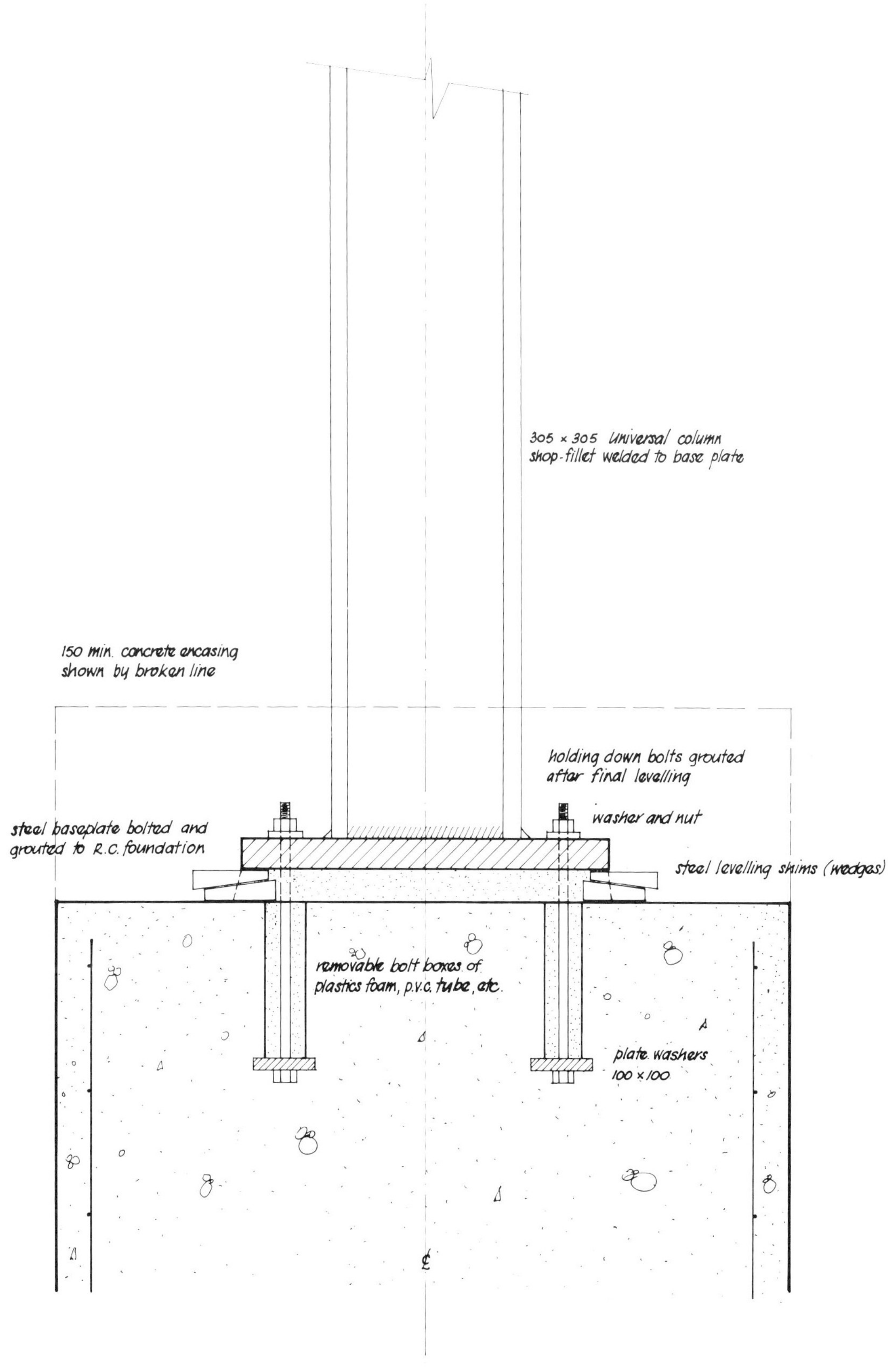

CONNECTION BETWEEN STEEL COLUMN AND CONCRETE BASE	NAME	D15/2
	DATE	
	SCALE 1 : 10	

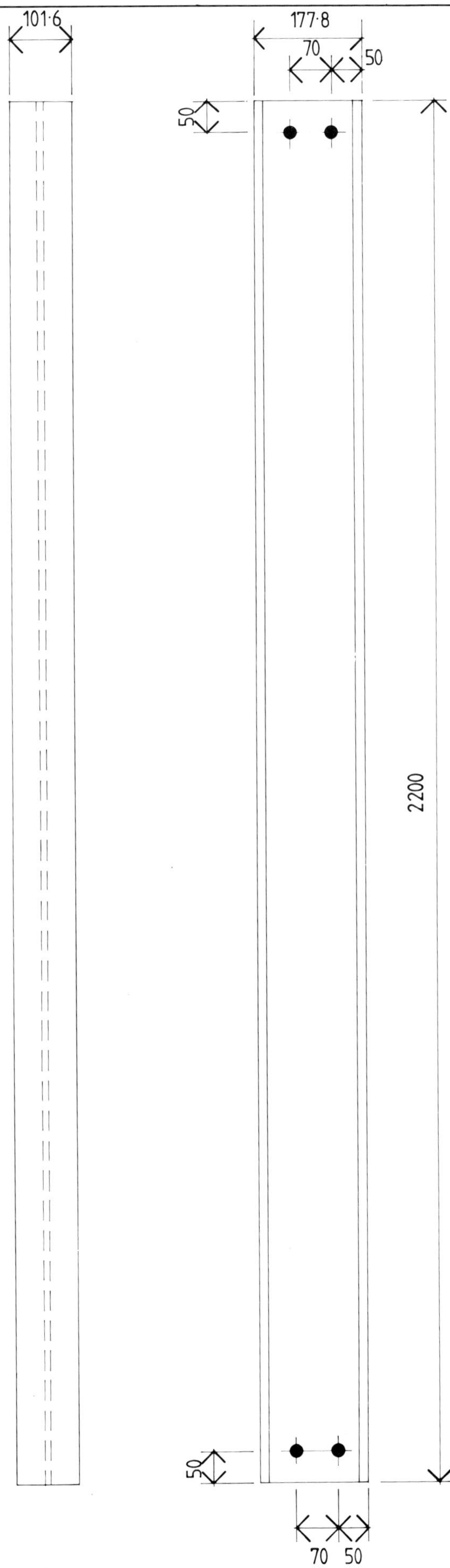

178 × 102 × 19 kg Universal Beam

All holes 22 mm diameter

DETAIL OF BEAM D-J14	NAME	D15/3
	DATE	
	SCALE 1 : 10	

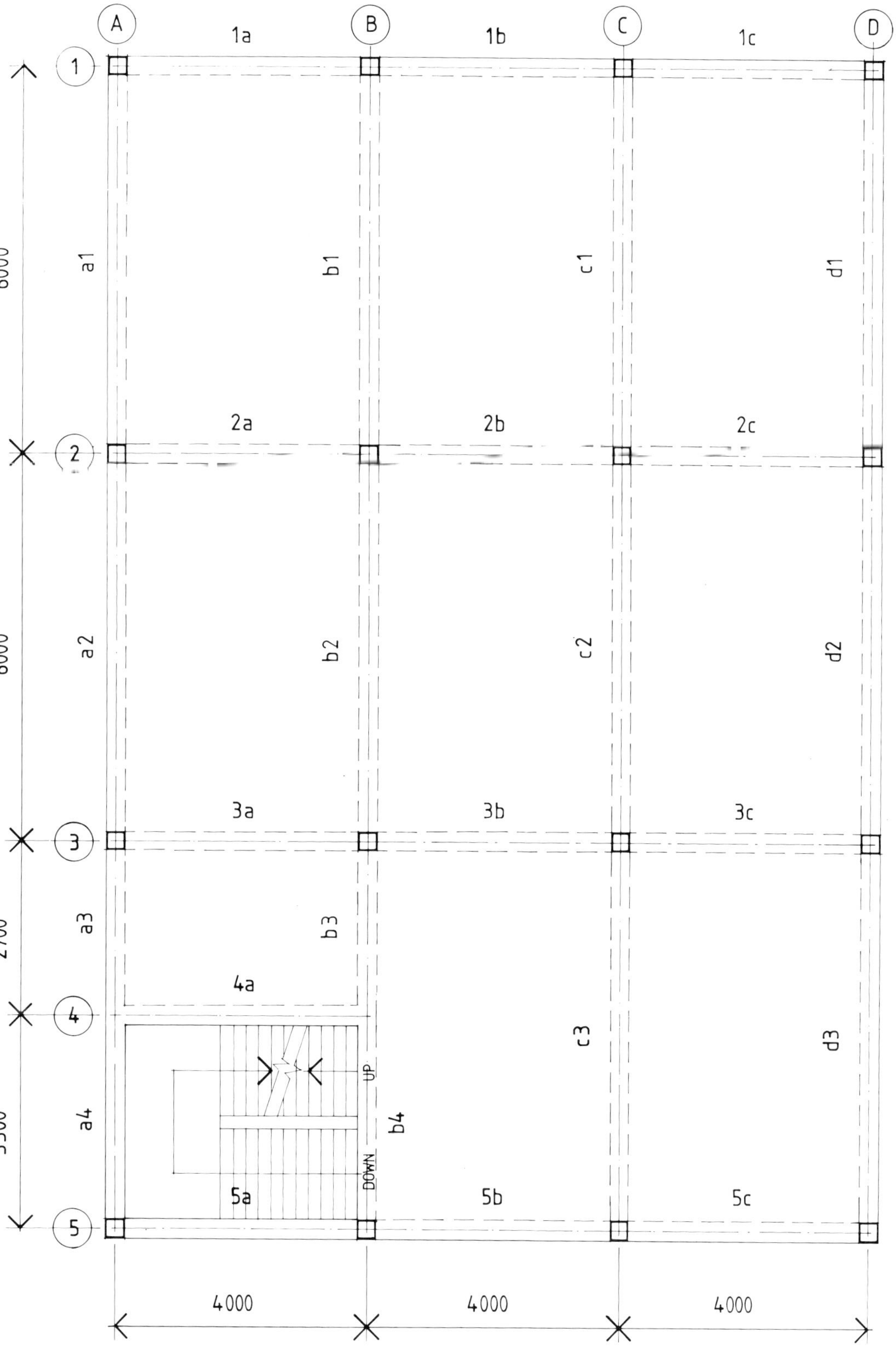

PLAN OF FIRST FLOOR SLAB	NAME	D15/4
	DATE	
	SCALE 1:100	

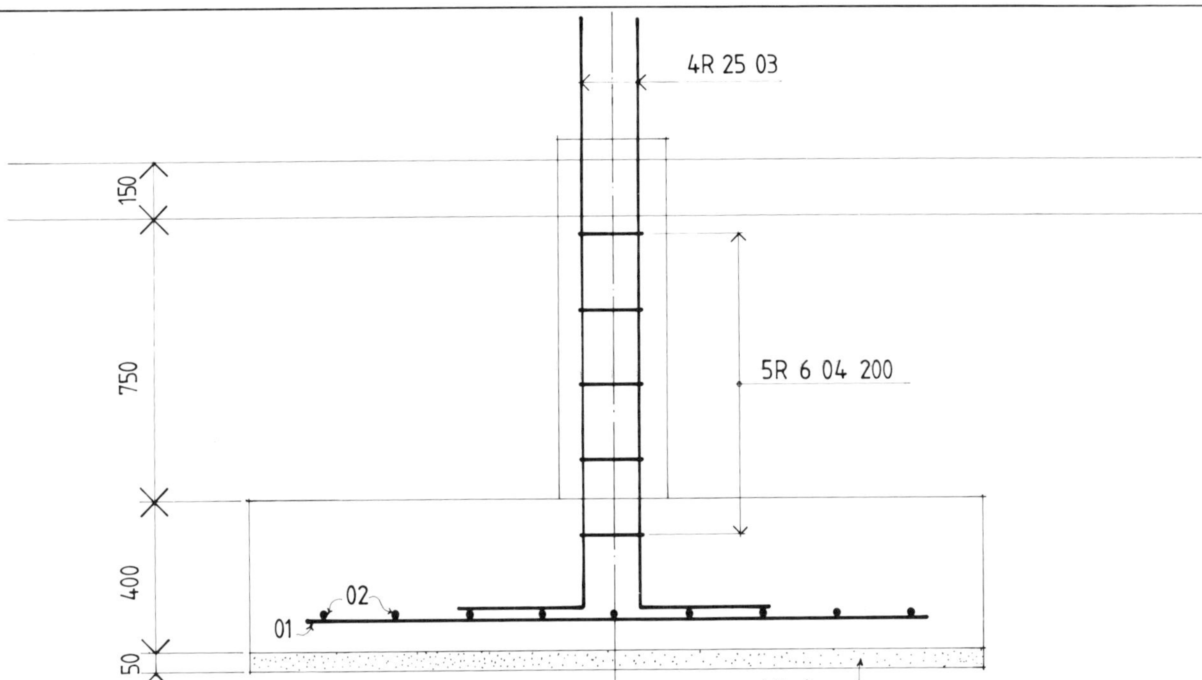

ELEVATION

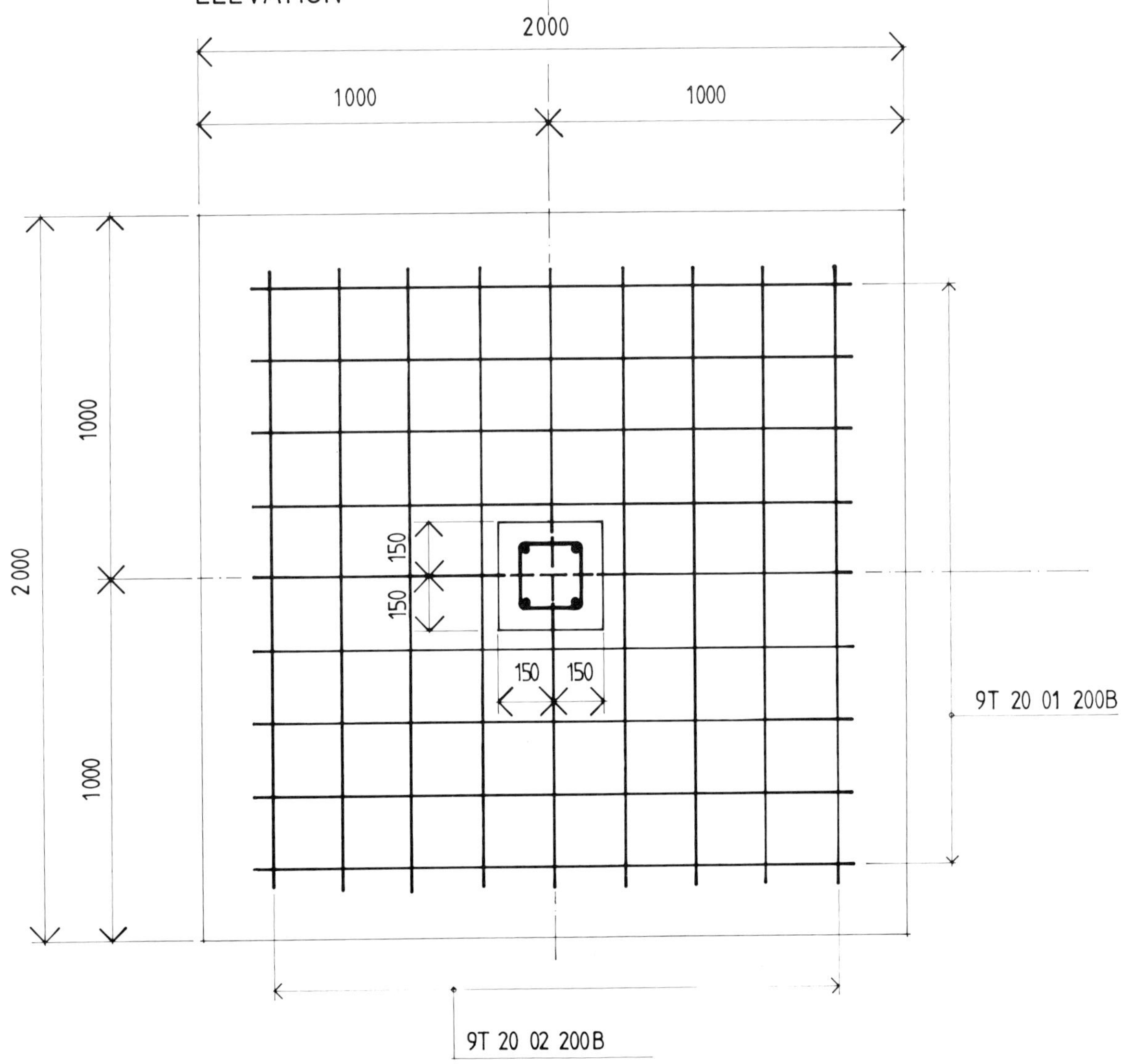

PLAN

R.C. COLUMN BASE

NAME

DATE

SCALE 1:20

D15/5

Dimensions

All setting out dimensions and dimensions of members (e.g. beams and columns) should be given on general arrangement drawings. The detail drawings need only show dimensions necessary to locate the reinforcement.

Dimension lines should be drawn outside the member to which they refer.

Exercise 15.1 Steel framing plan

Requirement
Copy drawing no. D15/1. Scale is to be 1:100.

Preliminaries
Undertake the preliminaries as described for previous exercises.

Construction lines
Draw the horizontal and vertical grid lines.

By scaling dimensions from drawing no. D15/1 draw the plan view of the columns.

Draw the horizontal and vertical dimension lines.

Draw the guide lines for the lettering and figuring.

Completion
Complete the drawing by lining in and lettering, as described for previous exercises. Use thick lines to define the beams and columns and thin lines elsewhere.

Exercise 15.2 Steel column connection to base

Requirement
Copy drawing no. D15/2. Scale is to be 1:10.

Preliminaries
Undertake the preliminaries as described for previous exercises.

Construction lines
Draw the horizontal line representing the top of the concrete base.

Draw the column vertical centre line.

Draw the two horizontal lines of the base plate.

Draw the four vertical lines of the column.

Draw the vertical/inclined lines of the base plate and fine concrete bedding.

Draw the bolts and boxes.

Draw the guide lines for the lettering.

Completion
Complete the drawing by lining in and lettering. Use thin lines throughout.

Exercise 15.3 Beam detail

Requirement
Copy drawing no. D15/3. Scale is to be 1:10.

Preliminaries
Undertake the preliminaries as described for previous exercises.

Construction lines
Draw the base line.

Draw the vertical lines, beginning with the left hand line.

Draw the top horizontal line.

Draw the bolt holes.

Draw the guide lines for the lettering and dimensions.

Completion
Complete the drawing by lining in and lettering. Use thin lines throughout.

Exercise 15.4 RC floor slab

Requirement
Copy drawing no. D15/4. Scale is to be 1:100.

Preliminaries
Undertake the preliminaries as described for previous exercises.

Construction lines
Draw the horizontal and vertical grid lines.

Draw the horizontal beam lines. All beams are 250 mm wide—125 mm each side of the grid line.

Draw the vertical beam lines.

By scaling dimensions from drawing no. D15/4 draw the staircase in the bottom left hand corner of the plan.

Draw the horizontal and vertical dimension lines.

Draw the guide lines for the lettering and dimensions.

Completion
Complete the drawing by lining in and lettering. Use thick lines for the columns and thin lines elsewhere.

Exercise 15.5 RC column base detail

Requirement
Copy drawing no. D15/5. Scale is to be 1:20.

Preliminaries
Undertake the preliminaries as described for previous exercises.

Construction lines
Draw the horizontal and vertical grid lines to the plan and elevation.

Draw the plan of the base by measuring and marking a distance of 1000 mm either side of both the horizontal and vertical grid lines.

Draw the plan of the column by measuring and marking a distance of 150 mm either side of both the horizontal and vertical grid lines.

Project vertical lines up from the plan to the elevation to define the sides of the column and base.

Draw the horizontal lines to the base and floor slab.

By scaling from drawing no. D15/5 draw the position of all the reinforcement bars on both the plan and elevation.

Draw the horizontal and vertical dimension lines.

Draw the guide lines for the lettering and dimensions.

Completion
Complete the drawing by lining in and lettering. Use thick lines to define the reinforcement and the plan view of the column and thin lines elsewhere.

Self Assessment Guide

In order to satisfy yourself that you have completed the exercises to an acceptable standard, study your drawings and check that you can answer YES to the following questions.

Exercise 15.1 Steel framing plan

1 Does your drawing look exactly the same as drawing no. D15/1 in this book?
2 Is your drawing accurately drawn to a scale of 1:100?
3 Have thick lines been used to define all columns and thin lines elsewhere?

Exercise 15.2 Steel column connection to base

1 Does your drawing look exactly the same as drawing no. D15/2 in this book?
2 Is your drawing accurately drawn to a scale of 1:10?
3 Have thin lines been used throughout?

Exercise 15.3 Beam detail

1 Does your drawing look exactly the same as drawing no. D15/3 in this book?
2 Is your drawing accurately drawn to a scale of 1:10?
3 Have thin lines been used throughout?

Exercise 15.4 RC floor slab

1 Does your drawing look exactly the same as drawing no. D15/4 in this book?
2 Is your drawing accurately drawn to a scale of 1:100?
3 Have thick lines been used for the columns and thin lines elsewhere?

Exercise 15.5 RC column base detail

1 Does your drawing look exactly the same as drawing no. D15/5 in this book?
2 Is your drawing accurately drawn to a scale of 1:20?
3 Have thick lines been used to define all the reinforcement and thin lines elsewhere?

Now check with each of your drawings that you can also answer YES to all the questions listed in the 'Standard Self Assessment Questions' at the back of the book (page 48).

16

Services Drawings

Types of Building Services Work

Building services are generally assumed to include hot and cold water supplies, above and below ground drainage, including sanitary appliances, refuse disposal, heating, ventilation, air conditioning, electrical installations including lighting, telecommunications, gas installations, fire protection, mechanical conveyors and security systems. Drawings are required for all of these services.

In this introduction to construction drawing a few simple examples are given of below ground drainage, water supplies and electrical installations for domestic buildings.

Main Groups of Drawings

The three main groups of drawings for illustrating services work are general layouts locating the arrangement of pipes, cables and ducts; details of plant areas; and details of specific items.

In addition there are schedules for items such as manholes, radiators, valves etc., but these are beyond the scope of this book.

Use of Grids

In projects where a structural grid is used, the services elements should be related to this grid.

In other cases the plant and equipment will be shown on services drawings in relation to a modular grid.

Identification of Services Components and Equipment

Services drawings provide information about a wide variety of different components and equipment. In order to identify individual items, it is usual to give them a reference number on drawings such as services layout drawings and location plans. An example is that a radiator might be given a reference of R305, which would mean it was a radiator, on the third floor, and was the fifth consecutive radiator on that floor.

General Layouts

Copy negatives (transparent copies) of the architect's 1:100 location drawings are often used by the building services engineers and technicians to show the general layout of the pipework, ductwork, trunking, cables etc. It is advisable to obtain these copy negatives from the architect at an early stage before too much detail is added. However, where this procedure is adopted, it is important that later revisions to the architect's drawings are taken into account on the copy negative.

A simple example of a general layout is shown on drawing no. D16/4, which is a wiring layout for lighting in a bungalow.

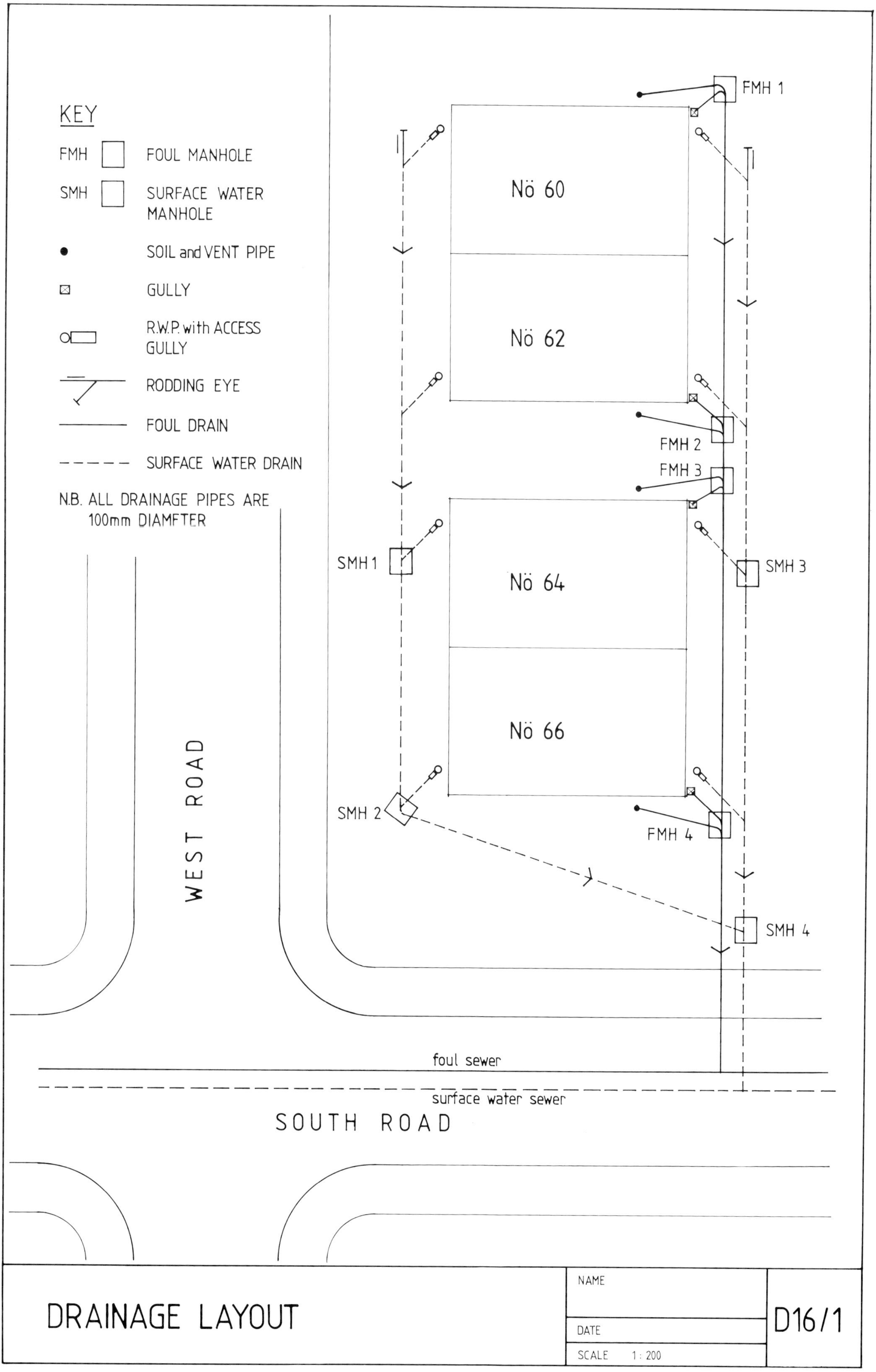

DRAINAGE LAYOUT	NAME	D16/1
	DATE	
	SCALE 1 : 200	

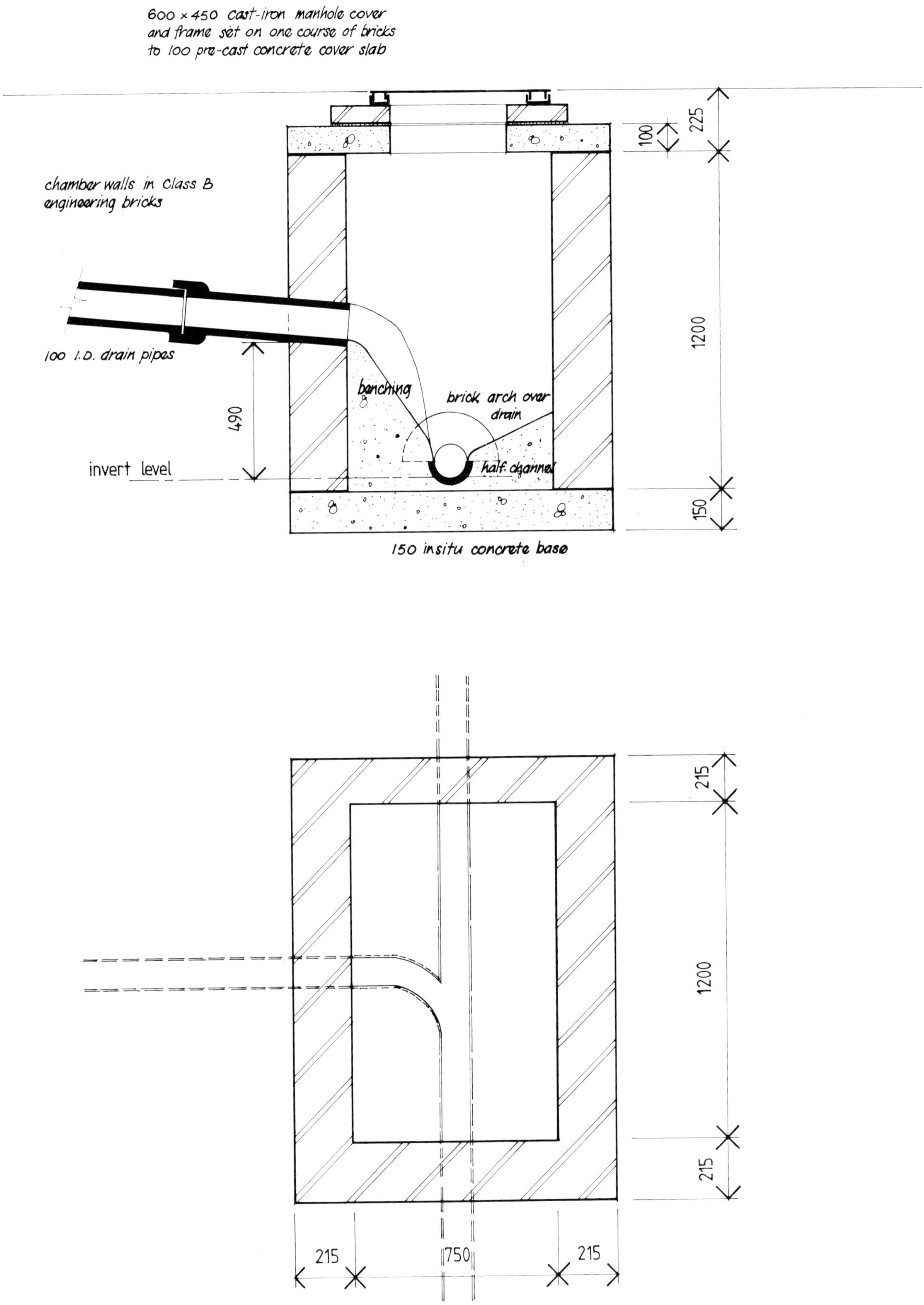

MANHOLE DETAIL

NAME

DATE

SCALE 1:20

D16/2

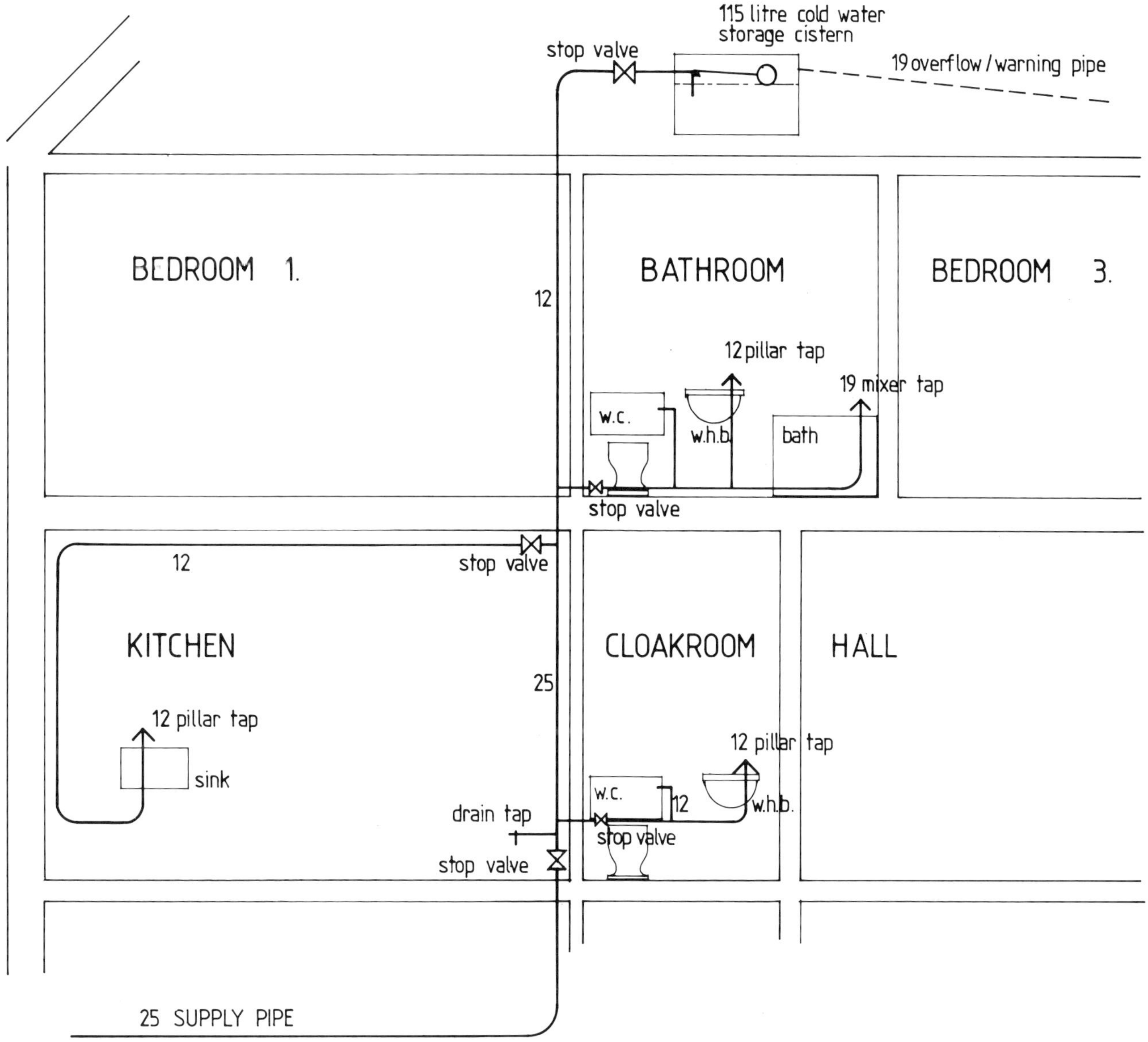

COLD WATER SUPPLY	NAME DATE SCALE 1:50	D16/3

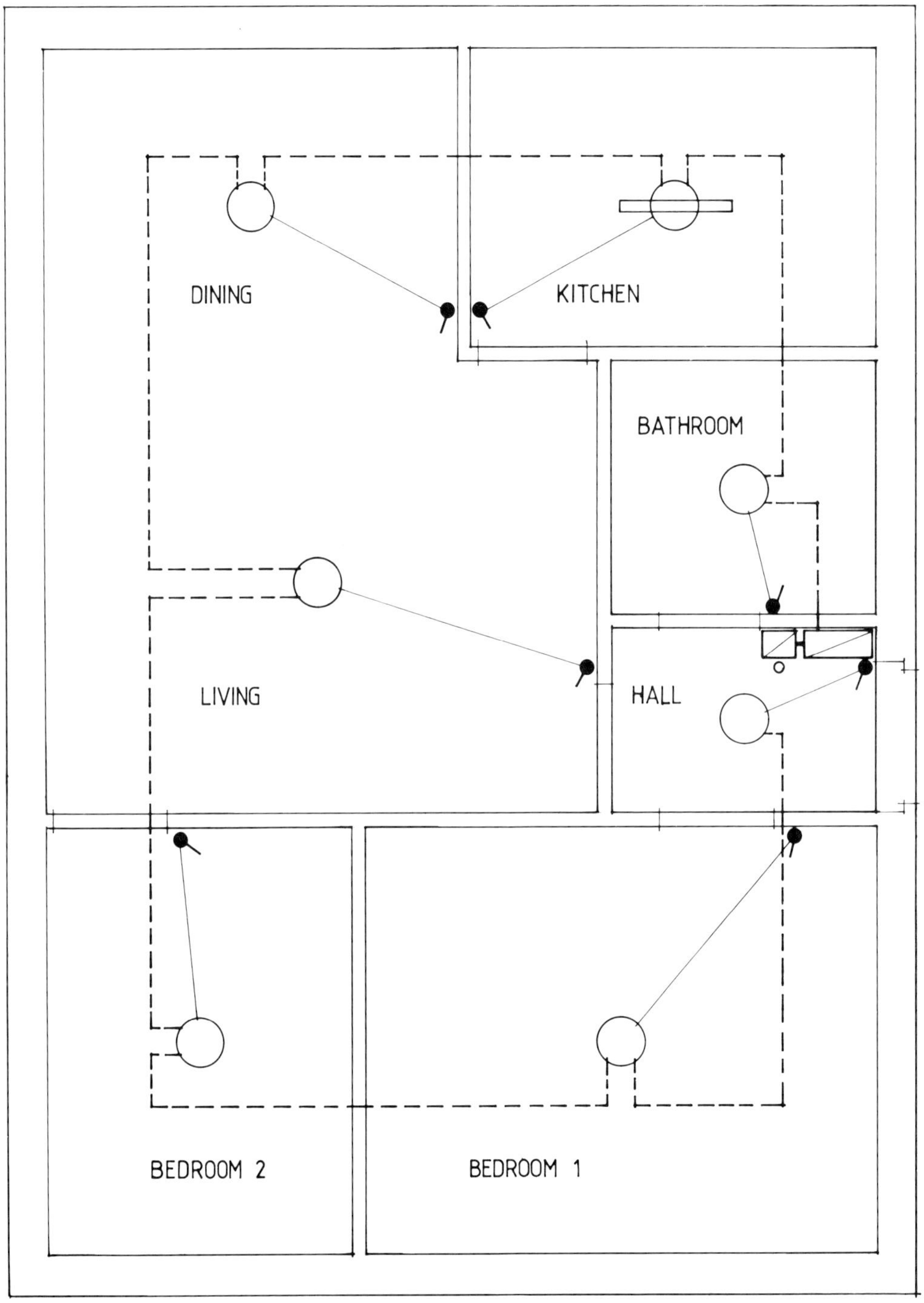

LIGHTING WIRING LAYOUT	NAME DATE SCALE 1:50	D16/4

Plant Area Drawings

Areas where the services equipment is concentrated are normally drawn to a larger scale, such as 1:50 and 1:20. Plans of these plant areas are the commonest form of plant area drawings, but elevations and sections are often required. Typical areas to be covered are boiler rooms, air handling plant rooms and electrical substations.

Details of Specific Services Items

As the scales of general layouts and plant room area drawings are comparatively small, additional detailed information on individual items is also required. This additional information is given on details of specific services items.

Drawing no. D15/2 is an example of details of a manhole.

Exercise 16.1 Drainage layout

Requirement
Copy drawing no. D16/1. Scale is to be 1:200.

Preliminaries
Undertake the preliminaries as described for previous exercises.

Construction lines
It is suggested that the various parts of the drawing be drawn in the following order,

Horizontal and vertical lines of houses.

Horizontal, vertical and curved lines to South Road and West Road.

Lines of drains and sewers.

Manholes, gulleys and other fittings.

Draw guide lines for lettering.

Completion
Complete the drawing by lining in and lettering, as described for previous exercises. Use thick lines to define drain and sewer lines and thin lines elsewhere.

Exercise 16.2 Manhole detail

Requirement
Copy drawing no. D16/2. Scale is to be 1:20.

Preliminaries
Undertake the preliminaries as described for previous exercises.

Construction lines
It is suggested that the various parts of the drawing be drawn in the following order.

Horizontal and vertical lines of the manhole walls on plan.

Horizontal, vertical and curved drain lines on plan.

Horizontal and vertical lines of concrete base on section.

Vertical lines of manhole walls on section.

Horizontal and vertical lines of cover slab, manhole cover and frame etc. on section.

Line of drain on section.

Lines of ramp and benching on section.

Dimension lines.

Hatching.

Guide lines for lettering and dimensions.

Completion
Complete the drawing by lining in and lettering, as described for previous exercises. Use thick lines for wall faces on plan and section, and for concrete faces on section; use thin lines elsewhere.

Exercise 16.3 Cold water supply

Requirement
Copy drawing no. D16/3. Scale is to be 1:50.

Preliminaries
Undertake the preliminaries as described for previous exercises.

Construction lines
It is suggested that the various parts of the drawing be drawn in the following order.

Horizontal floor and roof lines and vertical wall lines.

Sink, basins, bath and WCs.

Storage cistern.

Pipe lines.

Fittings—e.g. stop valves.

Guide lines for lettering.

Completion
Complete the drawing by lining in and lettering. Use thick lines for pipe lines and thin lines elsewhere.

Exercise 16.4 Lighting wiring layout

Requirement
Copy drawing no. D16/4. Scale is to be 1:50.

Preliminaries
Undertake the preliminaries as described for previous exercises.

Construction lines
It is suggested that the various parts of the drawing be drawn in the following order.

Horizontal and vertical lines defining wall positions.

Lighting points, distribution board and meter.

Wiring lines.

Switches.

Guide lines for lettering.

Completion

Complete the drawing by lining in and lettering. Use thick lines for the wiring lines and thin lines elsewhere.

Self Assessment Guide

In order to satisfy yourself that you have completed the exercises to an acceptable standard, study your drawings and check that you can answer YES to the following questions.

Exercise 16.1 Drainage layout

1 Does your drawing look exactly the same as drawing no. D16/1 in this book?
2 Is your drawing accurately drawn to a scale of 1:200?
3 Have thick lines been used for the drains and sewers and thin lines elsewhere?

Exercise 16.2 Manhole detail

1 Does your drawing look exactly the same as drawing no. D16/2 in this book?
2 Is your drawing accurately drawn to a scale of 1:20?
3 Have thick lines been used for the wall and concrete faces seen in section or sectional plan and thin lines elsewhere?

Exercise 16.3 Cold water supply

1 Does your drawing look exactly the same as drawing no. D16/3 in this book?
2 Is your drawing accurately drawn to a scale of 1:50?
3 Have thick lines been used for the pipework and thin lines elsewhere?

Exercise 16.4 Lighting wiring layout

1 Does your drawing look exactly the same as drawing no. D16/4 in this book?
2 Is your drawing accurately drawn to a scale of 1:50?
3 Have thick lines been used for the wiring and thin lines elsewhere?

Now check with each of your drawings that you can also answer YES to all the questions listed in the 'Standard Self Assessment Questions' at the back of this book (page 48).

17
Freehand Drawing

Use of Freehand Drawings

Freehand or sketch drawings are used for a variety of purposes. They may record or explain the appearance and construction of an existing building, or sketch in outline a designer's ideas for a proposed structure. Often freehand drawings will be used as preliminary constructional details, or to clarify on-site details which have not been made clear by the production drawings issued to the contractor. Freehand drawings may also be used as presentation drawings. This type will need to be of a high standard and will often incorporate advanced drawing techniques, including perspective drawing, shadow projection and rendering.

Drawing Techniques

The ability to produce quick, simple freehand sketches is a useful asset for anyone engaged in the construction industry. A picture will often describe and clarify things which are virtually impossible to explain with words.

In order to produce satisfactory freehand sketches of existing structures the draughtsperson will need to gain experience in the art of observation—sometimes referred to as training the eye. They will also need to acquire the ability to draw straight and curved lines of an even quality. Thirdly they will need to gain the ability to draw in proportion. These abilities will only be gained with practice, and some exercises to develop such skills are given in the next part of this chapter. Students are encouraged to use every opportunity to gain further practice, and it is suggested they carry a small sketch book and pencil with them when out of doors so that they can sketch any object of building interest which they see.

Production Information Sketch Drawings

Sometimes architects and other design team members will need to produce immediate information. The information must be precise and accurate, but can conveniently be provided in the form of freehand sketch drawings. An example of this may occur during the stage that the quantity surveyors are preparing their bills of quantities. A query may arise as to details of construction and finishes. The architect will explain to the quantity surveyor what is required, but it may be necessary to supplement the verbal description with a sketch. Examples of the type of sketch which would be produced in this situation are covered by Exercises 17.1, 2 and 3.

Some general advice on the matter of freehand sketches for production information is given below.

1 Draw everything first as light construction lines, and only firm in the lines when you are satisfied that everything is drawn accurately and in proportion.

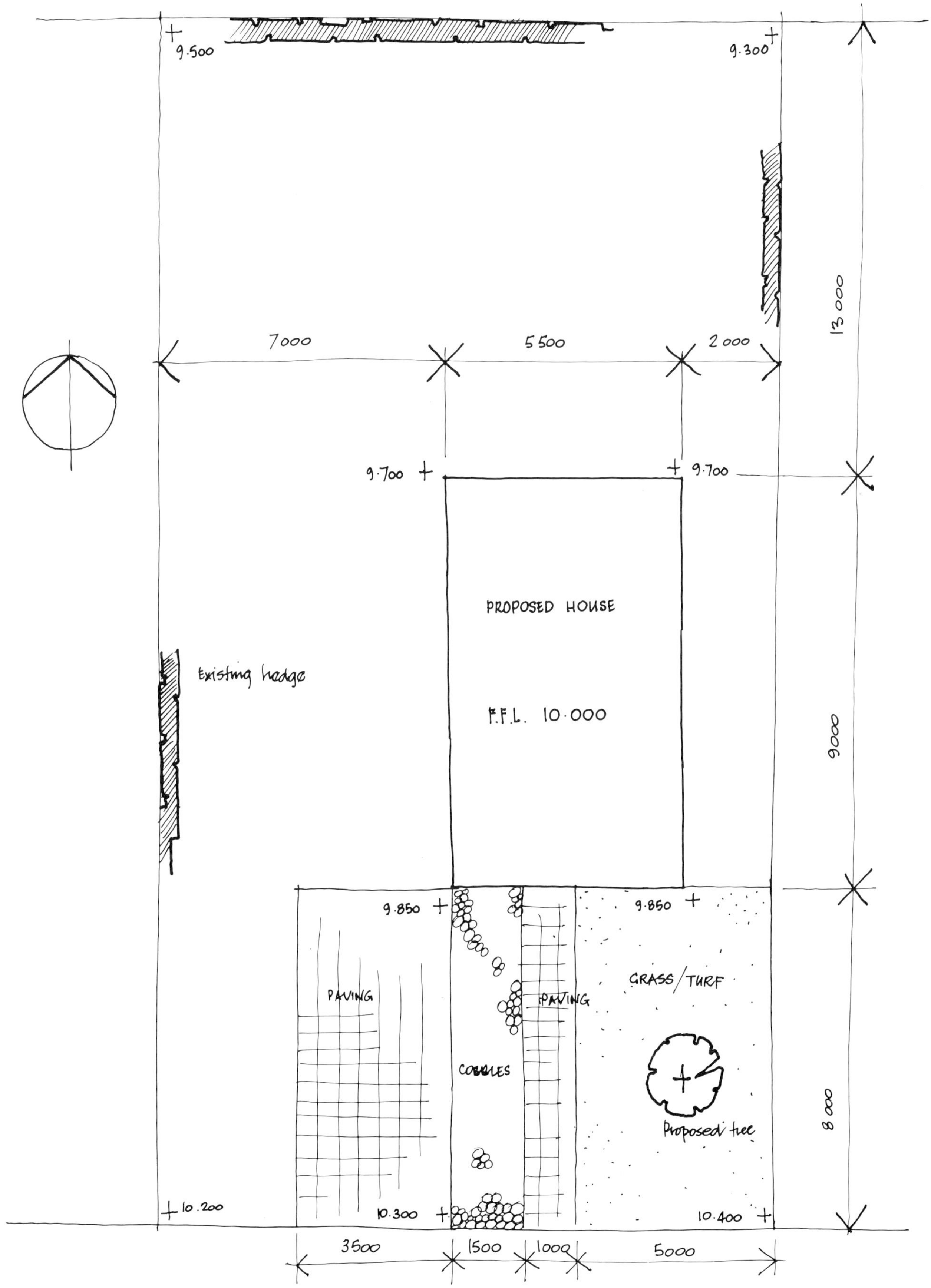

SKETCH DRAWING - 1

NAME

DATE

SCALE N.T.S.

D17/1

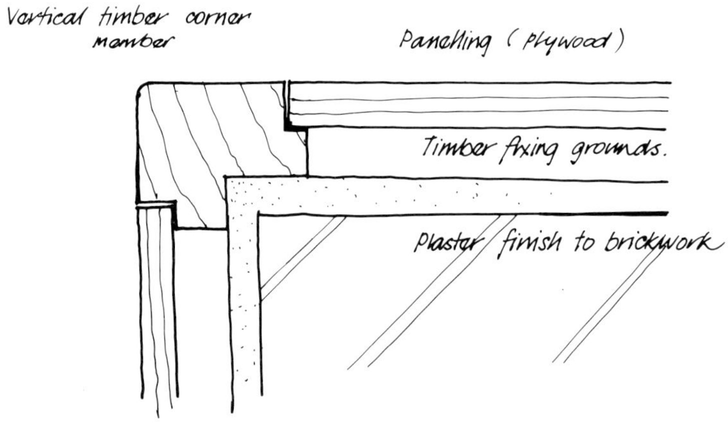

EXTERNAL ANGLE TO WALL PANELLING

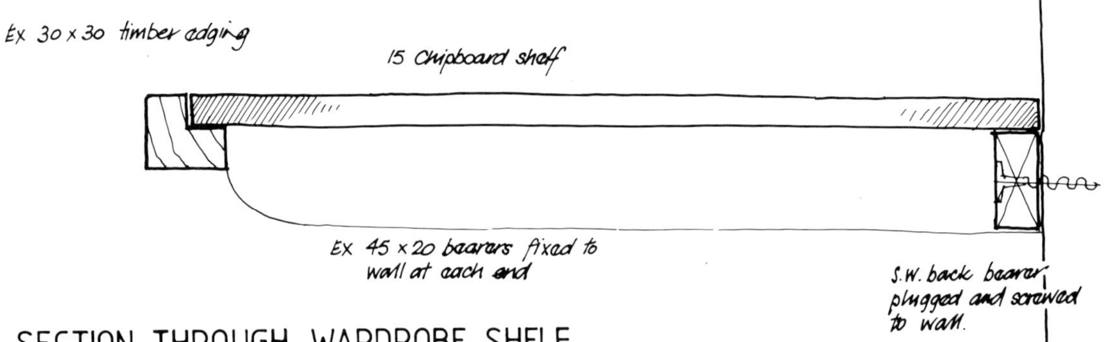

SECTION THROUGH WARDROBE SHELF

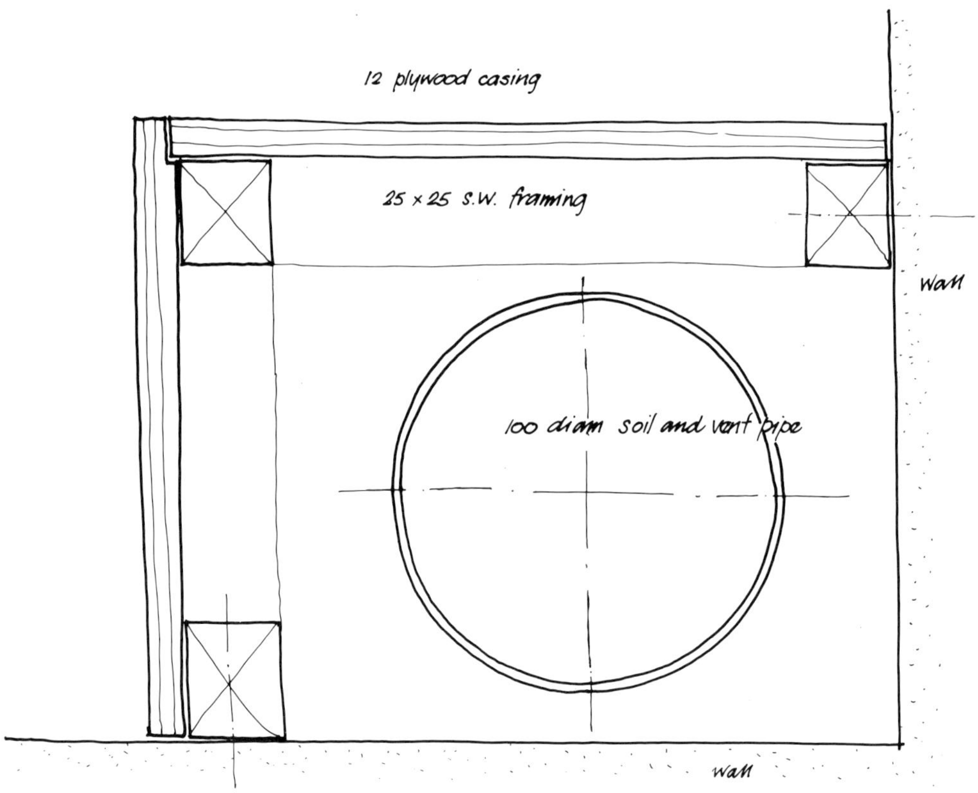

PLAN OF PIPE CASING

SKETCH DRAWING - 2	NAME	D17/2
	DATE	
	SCALE N.T.S.	

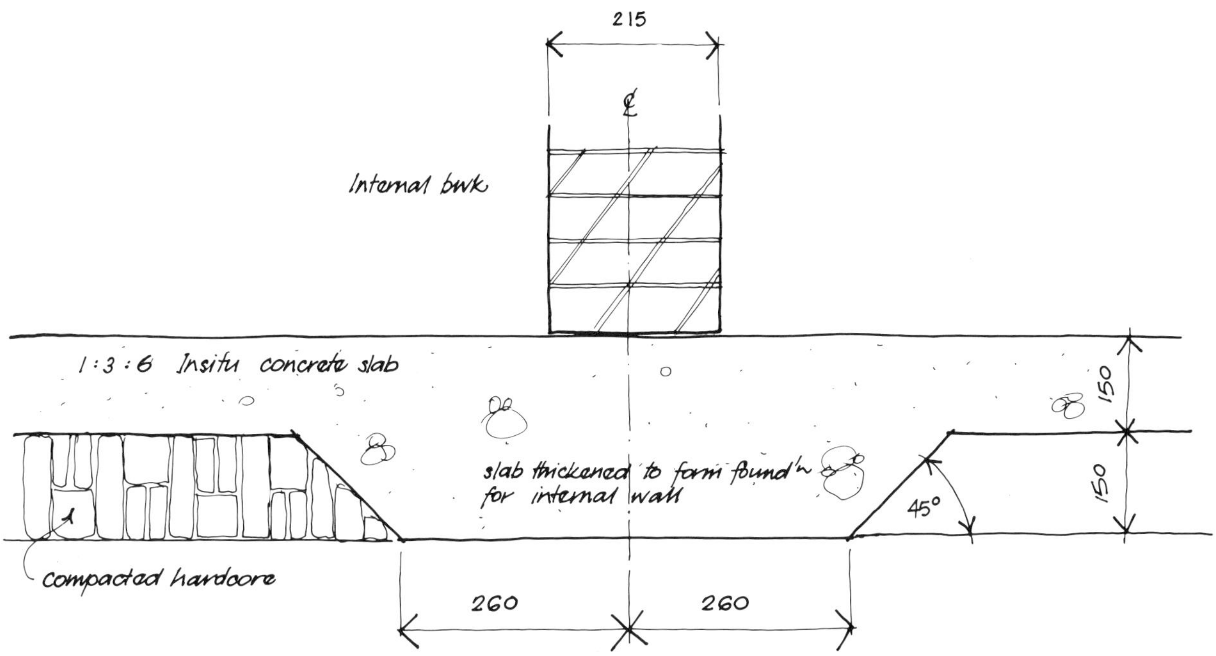

SECTION THROUGH BASE OF INTERNAL WALL

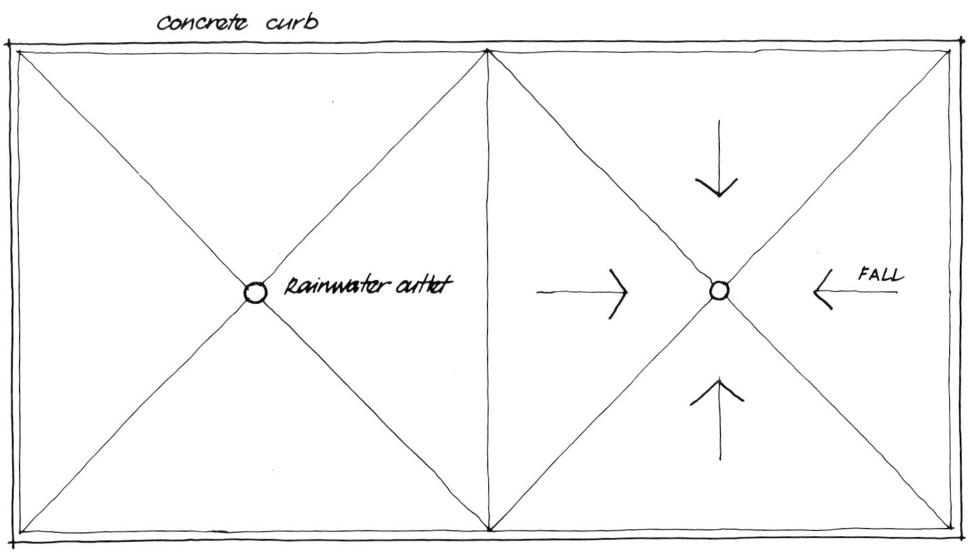

ROOF PLAN

SKETCH DRAWING - 3	NAME DATE SCALE N.T.S.	D17/3

2 Where feasible, divide what you are drawing into a number of simple geometrical shapes.
3 If what you are drawing is symmetrical, draw in the axes.
4 Draw in the main geometrical shapes first. Then add the detail.
5 Draw horizontal lines from left to right, unless you are left handed, in which case you will probably find it easier to draw them from right to left. If the line to be drawn is a long one, you can draw it as a continuous line made up of a series of shorter lines about 50 mm long.
6 Draw vertical lines from top to bottom.
7 Ensure that lines which are at right angles to each other are drawn as exact right angles.
8 In the case of circles first draw the axes, and mark the points on the axes where the circle is meant to cross.

Exercise 17.1 Production information, sketch drawing I

Requirement

Make a copy of drawing no. D17/1, in pencil, on a sheet of white A4 paper.

Preliminaries

Undertake the preliminaries as described for previous exercises. NB the lines to the borders and title panel are the only lines on the sheet which should be drawn with the aid of a straight edge/set-square.

Sketching

Begin by drawing all the lines as construction lines.

Draw a horizontal line fairly near the top of the sheet to represent the back boundary of the site.

Draw a horizontal line fairly near the bottom of the sheet to represent the front boundary of the site.

Draw two vertical lines, about 100 mm apart, to represent the side boundaries of the site.

About a quarter of the way up the right hand side (east) boundary of the site draw a horizontal line to represent the front face of the proposed house.

At a slightly greater distance draw a horizontal line to represent the back face of the house.

At a distance approximately half way across the width of the site draw a vertical line to represent the left hand (west) face of the house.

Now draw the right hand (east) face of the house. The gap between the east side of the house and the east boundary of the site is about one third the width of the house.

Now draw the lines defining the main areas of paving. The back line of this paving is a continuation of the line defining the front face of the house. The right hand side line is a continuation of the line defining the left hand (west) face of the house. The left hand side of the paving is defined by a vertical line midway between the left hand (west) boundary of the site and the left hand (west) face of the house.

Draw two further vertical lines to define the strip of cobbles and the path leading up to the front of the house. The width of the cobbles is a little less than half the width of the main area of paving, and the width of the path is two thirds the width of the strip of cobbles.

Draw in the proposed tree roughly in the centre of the grass turfed area.

Draw the dimension lines.

Check that everything has been included, and that the various parts are correctly proportioned.

Now firm in all the lines.

Add the hatching etc.

Add the dimensions.

Add the lettering.

Exercise 17.2

Production information, sketch drawing 2

Requirement
Make a copy of drawing no. D17/2, in pencil, on a sheet of white A4 paper.

Preliminaries
Undertake the preliminaries as described for previous exercises.

Sketching
Begin by drawing the detail at the top of the sheet. Initially all lines are to be drawn as construction lines.

Draw the external face of the wall. Start with the horizontal line. Then add the shorter vertical line.

Draw the horizontal and vertical lines representing the faces of the plaster finish.

Draw the horizontal and vertical lines representing the outside faces of the timber fixing grounds.

Copy the shape of the rebated vertical timber member. Add the panelling.

Check that everything has been included, and that the various parts are correctly proportioned.

Firm in all the lines.

Add the hatching.

Add the dimensions.

Add the lettering.

Next draw the detail in the centre of the sheet.

First draw three horizontal lines representing the shelf and the bearers. The distance between the two lines representing the bearers is three times the distance between the two lines representing the shelf.

Draw a vertical line representing the face of the wall.

Draw a vertical line to indicate the back bearer.

Draw a vertical line to define the front end of the bearers.

Draw the quarter circle rounding off the end of the bearer.

Draw the timber edging.

Check that everything has been included, and that the various parts are correctly proportioned.

Firm in all the lines.

Add the hatching.

Add the lettering.

Now draw the detail at the bottom of the sheet.

Draw the horizontal and vertical lines representing the wall faces.

Draw the axes of the circle representing the 100 mm soil and vent pipe. Mark the four points where the circle will pass through the axes, and then draw the circle.

Draw the two horizontal and the two vertical lines representing the two faces of the softwood framing. They should be spaced at about one third of the pipe diameter.

Draw the horizontal and vertical lines representing the outside face of the plywood. The thickness of the plywood is a little less than half the thickness of the softwood framing.

Check that everything has been included, and that the various parts are correctly proportioned.

Firm in all the lines.

Add the hatching.

Add the dimensions.

Add the lettering.

Exercise 17.3 Production information, sketch drawing 3

Requirement
Make a copy of drawing no. D17/3, in pencil, on a sheet of white A4 paper.

Preliminaries
Undertake the preliminaries as described for previous exercises.

Sketching
Begin by drawing the detail at the top of the sheet.

Draw the vertical chain dotted centre line in the position shown on drawing no. D17/3.

Draw the three horizontal lines representing the three horizontal surfaces of the concrete.

Mark off the two 260 dimensions along the lowest horizontal line of the concrete. These dimensions are slightly less than the maximum thickness of the concrete.

Draw the two short 45 degree lines joining the two bottom horizontal lines of the concrete.

Draw the vertical lines representing the two faces of the brick wall.

Draw in the brickwork mortar joints.

Check that everything has been included, and that the various parts are correctly proportioned.

Firm in all the lines.

Add the dimensions and lettering.

Now draw the detail at the bottom of the sheet.

Draw a square with sides about 60 mm towards the left hand side of the sheet. Make sure that all the sides are approximately the same length, and all the angles are right angles. Add a second square to the right of the first square. You now have a rectangle made up of two identical squares.

Draw lines just outside the two horizontal and the two vertical lines of the rectangle. These lines represent the outer face of the concrete curb.

Draw the two diagonal lines to each square. These lines represent the roof valleys. Draw the small circles in the centre of the squares. These represent the rainwater outlets.

Check that everything has been included, and that the various parts are correctly proportioned.

Firm in all the lines.

Add the arrows and lettering.

Self Assessment Guide

In order to satisfy yourself that you have completed the exercises to an acceptable standard, study your drawings and check that you can answer 'YES' to the following questions.

Exercise 17.1 Production information, sketch drawing 1

1 Does your drawing generally look the same as drawing no. D17/1 in the book?

2 Are the lines which are intended to be straight, reasonably true and even in quality?

3 Are the lines which are meant to be horizontal, perfectly true, and the lines which are meant to be vertical, perfectly true?

Exercise 17.2 Production information, sketch drawing 2

1 Does your drawing generally look the same as drawing no. D17/2 in the book?

Other questions are as for Exercise 17.1.

Exercise 17.3 Production information, sketch drawing 3

1 Does your drawing generally look the same as drawing no. D17/3 in the book?

Other questions are as for Exercise 17.1.

18

More Advanced Construction Drawing

Introduction

The drawing examples and exercises in this book are all of an elementary nature, mainly based on simple domestic structures.

Most drawings provide information about more complicated structures, although the principles outlined in previous chapters still apply. The purpose of this chapter is to make the student aware of additional drawings used by the construction industry, and the additional skills required by the experienced draughtsperson who prepares them.

Advanced Architectural Drawings

As buildings become larger resulting in more complicated forms of construction, with more stringent environmental and other requirements, the number, size, content and complexity of architectural drawings increases. Often there will be many floors in the building, with a separate drawing sheet needed for each location floor plan. There will also be several sheets of location sections and elevations, and numerous detail drawings illustrating the wide range of components and elements and the way they are assembled. The person preparing such drawings will need a considerable knowledge of construction, as well as an ability to organize the information presented on each sheet in a coherent fashion. Sets of drawings prepared for large projects need to be produced so as to achieve standardization of drawing techniques, annotation, dimensioning etc.

Some advanced architectural drawings, such as presentation drawings, require skills in perspective drawing, shadow projection, renderings in various media, transfer tones and colours, techniques of freehand drawing and additional lettering skills, including the use of stencils and dry transfers.

Sometimes there is a need to prepare measured drawings of buildings of historical interest, either for record purposes, or to meet town planning and other requirements when they are to be restored, altered or extended. People involved in the preparation of this type of drawing will often have to take accurate measurements of existing buildings, and carefully plot the results. Skill in preparing accurate freehand sketches, from small scale plans and elevations to full-size details, and including perspectives, will frequently be required.

Useful information on all the above types of advanced architectural drawings are given in an excellent book entitled *Draughtsmanship architectural and building graphics* by Fraser Reekie.

Advanced Structural Engineering Drawings

The preparation of structural engineering drawings also requires additional skills as the buildings become larger and more complex. Examples of more complicated work include structures such as basements and other substructures, shell roofs, including hyperbolic paraboloids, jack knife and spiral staircases, folded plate slabs and structures incorporating raking struts.

Advanced Services Drawings

As was mentioned in Chapter 16, drawings are required for a wide range of building engineering services. These include air conditioning drawings, including ductwork layouts and details of air handling plant rooms; heating systems, including pipework layout and details of the boiler room; and sprinkler installations.

In complex projects containing a wide range of services, co-ordination drawings must be produced to avoid conflict between the various services. It is important to show the interrelationship between different services, to ensure that several engineering services are not competing for the same space in plant rooms, service ducts and floor, ceiling and roof voids.

Co-ordination is also necessary between structural and architectural elements.

Landscape Drawings

Landscape drawings may need to be produced for larger projects. They are concerned with work related to the external environment, including the layout and planting of spaces between and around buildings.

Survey drawings and maps, including ordnance survey maps, often form the basis for landscape drawings, so the draughtsperson needs to be familiar with these mediums, as well as contours, topographical details and conventional representation for survey drawings.

Geometrical Drawing

Sometimes designers of buildings use geometrical shapes such as circles, ellipses and parabolas. Draughtspersons preparing drawings for such buildings will need to have a sound knowledge of geometrical drawing.

Examples of geometrical elements which can form parts of buildings are domed roofs, hyperbolic paraboloid roofs, curved niches, spiral staircases, parabolic arches and elliptical openings.

Computer Aided Draughting

Another area of more advanced construction drawing is computer aided design and draughting (CAD). In recent years increasing use has been made of computerised design and draughting systems, and prices have now reached a level where even the smaller architectural practices can afford a sophisticated micro system.

The use made of computers will vary from office to office. The term computer aided design is widely used, but a computer system which can design as well as draw, and has the capacity to produce both 2D and 3D graphical solutions from tabulated data is likely to be expensive and may be out of the reach of smaller firms. It is more usual for computers to be used as automated draughting systems. The design stage of a project is often carried out in a traditional way, and even when a perspective drawing is produced with the aid of a computer it may be finished off manually. Frequently the computer only comes into its own at the working (production) drawing stage.

Initially there may not seem to be a big advantage in using a CAD system for the preparation of working drawings. For example the saving in time by using a CAD system for producing a location floor plan as opposed to drawing it manually may not seem to be very dramatic. The real advantage comes when drawings have to be amended, or partly or wholly reproduced for use as structural and services drawings etc., as these operations can be done easily and quickly with CAD.

Self Assessment Guide

In order to check whether you have understood all you have read in this chapter, you should now attempt to answer the following questions.

1 State the extra skills required by persons preparing drawings for larger buildings as compared to small domestic buildings.
2 What are measured drawings?
3 Name two types of lettering which may be used on construction drawings other than freehand lettering.
4 Give two examples of structural work more complicated than that covered by this book.
5 What is the purpose of co-ordination drawings?
6 What are landscape drawings concerned with?
7 Why is it sometimes necessary to have a knowledge of geometrical drawing when preparing construction drawings?
8 Describe the advantages of CAD as compared to working drawings prepared manually.

When you have attempted all the above questions, reread the appropriate part of the chapter and check you know the correct answer to every question.

19

Organisation and Management of Construction Drawing Work

Decision Making

As projects become larger and more complicated it is necessary to adopt correct techniques for organizing and managing the drawing work. However even on small, simple projects many of the general principles outlined below will apply.

Whenever possible the following decisions should be taken before work on the production information drawings begin.

Drawings

- prepare a list giving the title of each drawing required
- decide the content of each drawing
- choose scales for each drawing
- decide the size of the standard drawing sheets
- determine the reference techniques to be followed, including the grid
- decide the tolerances to be allowed
- agree the style and content of the title panel

Outstanding information

- list all outstanding information
- record who is to supply this information
- agree dates when the information will be provided

Programme

- Agree a programme showing the dates each drawing is to be provided

Drawing Registers

A drawing register is an important part of management procedures. It can provide the following information about each drawing.

- number
- title
- discipline—e.g. architectural, structural engineering
- type—e.g. location drawing, assembly drawing
- scale
- sheet size—e.g. A1
- revision suffix
- who has been sent a copy

Drawing Numbers

Various numbering systems are used by construction design staff. A typical system will consist of the following components.

1 project number or letters
2 discipline letter—for example:
A for architectural

S for structural engineering
D for drainage
W for water services
H for heating
C for air conditioning
E for electrical
L for landscaping

3 identifying number within each discipline
4 revision suffix

A drawing with the number 359A/12a would mean the project has been given the number of 359; it is an architectural drawing and the twelfth such drawing to be prepared for the project, with the revision suffix of 'a' denoting there has been one revision since the drawing was first issued.

Checking Drawings

In order to reduce the risk of errors and omissions it is important to check every drawing before it is issued. The draughtsperson responsible for the drawing should make his or her own check, but whenever possible a further check should be made by someone else sufficiently experienced to recognize the accuracy and completeness of the drawing.

The procedure is to check that the drawing is

- technically correct and complies with all regulations
- dimensioned accurately, arithmetically correct and drawn to scale
- complete and includes all relevant information
- of an acceptable quality as regards lines, legibility and general appearance
- provided with the correct symbols and uses standard conventions
- annotated satisfactorily
- supplied with a correct title panel

Check Lists

Check lists are useful tools during both the preliminary stage of organizing the drawing work and the actual drawing stage. They help to ensure that a systematic approach is adopted in directing the drawing work, and are a means of highlighting decisions which have to be made and avoiding errors and omissions.

Examples of two such check lists are given below. The first of these deals with the organization of the production information, including the construction drawings. The second is intended to make certain that each drawing contains all the necessary information.

Check List for Organisation of Production Information

1.00 *List of drawings and schedules to be prepared*
N.B. Content of each drawing is to be listed with scales

1.01 Location drawings
a) Block plans
b) Site plans
c) Floor plans
d) Foundation plans
e) Roof plans
f) Sections
g) Elevations

1.02 Assembly drawings
a) Use of standard drawings. N.B. Quote detail numbers
External walls
Wall openings
Eaves details

Parapet details
Roof lights
Internal walls
Internal wall openings
Suspended ceilings
b) Non standard drawings. N.B. List each drawing

1.03 Component drawings
a) Use of standard drawings. N.B. Quote detail number
Windows
Doors
b) Non standard drawings. N.B. List each drawing

1.04 Schedules
a) Windows
b) Doors
c) Ironmongery
d) Manholes

1.05 Structural drawings
a) Structural steel general arrangements
b) Fixing details
c) Details of individual members
d) Reinforced concrete general arrangements
e) Reinforcement details
f) Profile details
g) Other drawings
h) Schedules

1.06 Building engineering services drawings
a) Drainage
b) Discharge pipework
c) Electrical
d) Water
e) Gas
f) Heating and air conditioning
g) Other drawings

2.00 *Drawing procedures*
2.01 Standard sheet sizes
2.02 Scales
2.03 Title panel/job title/job number/numbering system
2.04 Annotation style—stencil or freehand
2.05 Use of computers
2.06 Production of copy negatives

List who is to prepare and receive negatives

3.00 *Referencing*
3.01 Grid
3.02 Floor levels
3.03 Floor numbers/letters
3.04 Room numbers
3.05 Equipment numbering
3.06 North point

4.00 *Agree tolerances*
4.01 Structure
4.02 Services

5.00 *Specifications*
5.01 Method
5.02 Contributions from consultants
5.03 Contributions from subcontractors and suppliers

6.00 *Outstanding information required*
6.01 Client

6.02 Structural engineering consultants
6.03 Building engineering services consultants
6.04 Subcontractors and suppliers
6.05 Others
6.06 Agree timetable

Drawings Check List

1.00 *General*
1.01 All information applicable in title panel completed
1.02 Revision letters and descriptions
1.03 Cross references to other drawings and general notes
1.04 North point
1.05 Datum for levels

2.00 *Site and layout plans*
2.01 Setting out dimensions
2.02 Existing and new levels
2.03 Landscaping (e.g. trees to be retained or removed; fences and hedges to be removed; ditches filled in; pavings)
2.04 Building line
2.05 Fences, gates and new accesses
2.06 Public footpaths and rights of way
2.07 Adjacent buildings
2.08 Road names
2.09 Paths
2.10 Sewer mains and invert levels
2.11 Gas main
2.12 Electricity
2.13 Telephone
2.14 Water supply
2.15 Drainage

3.00 *Floor plans*
3.01 Dimensions (including wall thicknesses)
3.02 Floor areas
3.03 Floor levels
3.04 Section lines
3.05 Room names and/or numbers
3.06 Floor finishes
3.07 Beam over (dotted)
Rooflights over (dotted)
3.08 Services ducts
3.09 Flues
3.10 Vertical damp proof courses
3.11 Built in furniture
3.12 Door numbers; swings
3.13 Window numbers
3.14 Stairs; number treads, show directions up and/or down
3.15 Roof space access
3.16 Expansion joints

4.00 *Roof plans*
4.01 Dimensions
4.02 Falls
4.03 Gutters and falls
4.04 Rainwater outlets
4.05 Roof finishes
4.06 Parapet copings
4.07 Maximum and minimum screed thicknesses
4.08 Flues
4.09 Vents

4.10 Rooflights
4.11 Lightning conductors
4.12 Expansion joints

5.00 *Sections*
5.01 Structural and finished floor levels
5.02 Floor to ceiling (or floor to floor) heights
5.03 Vertical dimensions
5.04 Foundation dimensions
5.05 Ground level (existing and finished)
5.06 Damp proof courses and membranes
5.07 Roof pitches
5.08 Tanking

6.00 *Elevations*
6.01 Floor lines (related to datum)
6.02 Windows including opening types
6.03 Cill heights
6.04 Soil and vent pipes
6.05 Rainwater pipes and gutters
6.06 Rainwater heads
6.07 Expansion joints to brickwork etc.
6.08 Window numbers

7.00 *Foundation plans*
7.01 Dimensions; levels
7.02 Reinforcement to concrete
7.03 Position of walls
7.04 Damp proof courses
7.05 Expansion joints

8.00 *Drainage plans*
8.01 Sewer main (type, i.e. foul, combined etc.)
8.02 Position of sanitary fittings
8.03 Soil and waste pipe layout
8.04 Inspection chamber and manhole positions; sizes; inverts
8.05 Vent pipes
8.06 Gulleys
8.07 Grease traps
8.08 Rodding eyes
8.09 Cesspool
8.10 Soakaways
8.11 Septic tank
8.12 Rainwater pipes
8.13 Interceptors

9.00 *Electrical services*
9.01 Lighting points
9.02 Power points
9.03 Switches
9.04 Rising main
9.05 Meter cupboard
9.06 Bell installation
9.07 Alarm system

10.00 *Water services*
10.01 Rising main
10.02 Point of entry/first stop cock
10.03 Cold water storage tank
10.04 Hot water cylinder
10.05 Hot water expansion tank
10.06 Stop cocks
10.07 Taps

10.08 Sanitary fittings
10.09 Fire fighting equipment
10.10 Overflow pipes
10.11 Drain cocks

11.00 *Heating and air conditioning services*
11.01 Pipe work
11.02 Ductwork
11.03 Boilers
11.04 Tanks
11.05 Pumps
11.06 Terminal units
11.07 Air conditioning plant
11.08 Louvres

Self Assessment Guide

1 Before work on the production information drawings begin certain decisions and actions have to be taken. Summarize these under the following headings: (a) drawings; (b) outstanding information; and (c) programme.
2 What information can be provided in a drawing register?
3 In respect of drawing numbers, what is meant by the terms: (a) discipline letter; and (b) revision suffix?
4 Who should check a construction drawing before it is issued?
5 List the items which should be included in a drawing check.

When you have attempted all the above questions, reread the appropriate part of the chapter and check you know the correct answer to every question.

20
Words and Phrases

This chapter provides a summary, in alphabetical order, of the meaning of the various words and phrases used in the subject of construction drawing.

Annotation
Notes added to a drawing.

Assembly drawing
A drawing which shows how the parts of a building are constructed, and how they meet at junctions.

Axonometric projection
A method of showing an object or building in three dimensions, and incorporating a true plan. The view is usually taken looking down at an angle on the object of building.

Block plan
A plan which identifies a site, and locates the outline of the buildings on that site in relation to a wider context such as a town plan.

British Standards Specification
These are issued by the British Standards Institution to lay down minimum standards for materials and components, including drawing instruments such as scale rules and set-squares.

Building line
A line set by the local authority on or behind which buildings must lie.

Cartography
Map making.

CI/SfB
A system for filing technical information about the construction industry.

Civil engineering
The design and construction of structures such as roads, bridges, tunnels, harbours, dams, airstrips, irrigation systems and similar unroofed structures.

Codes of practice
Documents issued by the British Standards Institution to cover design and workmanship in specific areas, including draughtsmanship.

Component drawing
A drawing which shows the shape, dimensions and other details of components, such as windows.

Construction drawings
Drawings prepared to show builders and other involved parties the shape, size and position of buildings on site, together with the way the buildings and equipment are constructed.

Construction lines
Faint guide lines, or setting out lines, drawn in the initial stages of preparing a drawing. In the later stages they are either covered by final lines, or erased.

Contract drawing
A drawing illustrating work forming part of a contract, which is legally binding upon the parties to that contract.

Co-ordination drawing
A drawing showing the interrelationship between the structure, architectural features, engineering services and other constituents of a building.

Datum
A fixed horizontal base line or level from which heights and depths can be measured.

Detail drawing
A drawing illustrating components, elements, or parts of the building, generally to a larger scale, and providing more detailed information.

Development plan
Plan prepared by the local district or county council to protect the environment, and achieve a balanced allocation of the available land to meet the various needs. Among other things it stipulates the type of development (houses, factories etc.) permitted in a particular area.

Dimension
A measure in length, width, or height.

Dimensioning
The addition of dimensions to a drawing.

Drawing
Information depicted graphically.

Elevation
View on a vertical plane, particularly a view of one side of a building.

Final lines
Lines which are part of a completed drawing. They are sometimes called finished lines.

General arrangement drawing
A drawing providing general information about construction work, such as the general layout of beams and columns, or the general layout of engineering services.

Grid
A network of horizontal and vertical reference lines.

Isometric projection
A method of showing an object or building in three dimensions, but distorting the plan view so as to achieve a more realistic effect.

Location drawing
A drawing locating and identifying spaces and parts of a building, such as rooms, drainage and doors. Examples of location drawings are 1:100 plans, elevations and sections.

Oblique projection
A method of showing an object or building in three dimensions, and incorporating a true elevation.

Orthographic projection
A form of drawing in which every part is drawn as if immediately in front of the observer, using two dimensional views.

Plan
A view of a building or object on a horizontal plane, commonly the view obtained by looking down on a building or object.

Planometric projection
Another name for axonometric projection.

Production drawing
A drawing used to assist the manufacture of components, and the assembly of materials and components on a construction site. The main types are location, assembly and component drawings.

Production information
The total information, including drawings, schedules and specifications, produced by the design team to facilitate the erection of a building.

Record drawing
A drawing which records the way 'construction works' or part of those works have been constructed.

Scale
The ratio of size between an object or building illustrated on a drawing and the actual size of that object or building.

Schedule
A document giving tabulated information about similar items such as reinforcement bars, manholes and doors.

Section
A view of an object or building obtained by making an imaginary cut through it, usually on a vertical plane.

Site plan
A plan locating the position on a site of buildings and other construction works in relation to reference points such as boundaries and accesses.

Sketch drawing
Preliminary drawing indicating the designer's general intentions.

Working drawing
An alternative name for a production drawing.

21

Exercises and Self Assessment Questions

Additional Exercises and Assignments

After completing all the exercises at the end of each chapter, it is suggested that students attempt the following additional exercises and assignments.

1 Draw drawing no. D17/1 to a scale of 1:200.
2 Draw the site plan shown on drawing no. D9/1 to a scale of 1:100, on a sheet of A2 paper.
3 Draw the longitudinal section of the bungalow shown on drawing numbers D10/1, D11/1 and D12/1 to 12/4 to a scale of 1:50, on a sheet of A3 paper.
4 Prepare a production drawing of the bungalow shown on numbers D9/1, D10/1 D11/1 and D12/1 to 12/4. The floor plan, cross section, longitudinal section and four elevations should be drawn to a scale of 1:100 and the site plan to a scale of 1:200.
5 Prepare, without the use of scales, set squares or any other drawing aids, freehand drawings of drawing numbers D9/1, D10/1, D11/1 and D12/1.
6 Prepare, without the use of scales, set squares or any other drawing aids, freehand drawings of the assembly drawing numbers D13/1 to 13/4.
7 Undertake the following assignments based on your own home, or the home of a relative or friend. The scales used are to be at your discretion.

See 'Brief information on the preparation of measured drawings' on page 147.

7(a) Measure up the home and produce scale drawings of (i) floor plans; and (ii) front and back elevations.
7(b) Prepare a tracing of the floor plan and add the position of the following on the plan.

- radiators
- electric socket outlets
- lighting points and switches
- hot and cold water taps.

7(c) Produce an axonometric drawing of the kitchen, including all fixtures and fittings.
7(d) Produce an isometric sketch of the probable runs of the hot and cold water pipes.
7(e) Sketch and measure the doors in the home and prepare component drawings of these doors.

7(f) Sketch and measure the windows and prepare component drawings of these windows.

7(g) Prepare a drawing detailing the route of the waste water from the kitchen sink to the local authority sewer. This is to be done by making sketches where the route is visible and making assumptions elsewhere.

Brief Information on the Preparation of Measured Drawings

1. Procedures

The work should be completed in three stages, as follows.

(a) Walk round the building for which a measured drawing is to be prepared, and produce freehand sketch drawings.

(b) Take measurements and add them to the sketch drawings.

(c) Produce final drawings to scale, using instruments—tee squares, set squares etc.

2. Sketch drawings

(a) Carefully draw sketch drawings, in proportion, making sure that everything, particularly the dimensions, are neat and easy to understand. The golden rule is that someone who has never seen the building should be able to draw an accurate final drawing of the building, using only the information contained on your sketch drawings. Guidance on freehand drawing was given in Chapter 17.

(b) Make separate details of any special features.

3. Measurements

(a) Plans

Wherever possible take running dimensions, with the dimensions running from left to right (see Fig. 15.2A).

When it is not possible to take running dimensions, individual dimensions should be taken (see Fig. 15.2A).

Take each full room dimension, (i.e. overall length and width of the room) and compare it with the sum of individual dimensions, made up for example of the width of door opening, chimney breast, alcove etc.

Take plenty of dimensions and where possible relate them to each other.

Dimensions of diagonals should be taken to check whether rooms are square at corners.

Remember to measure the thickness of all walls. These can generally be taken at door and window openings.

(b) Elevations

With brick buildings approximate dimensions of inaccessible areas can be obtained by counting brick courses. Typically four courses of bricks measure 300 mm. Elevations in stonework or render are more difficult. Accessible parts of the building can be measured, and dimensions of other areas can then be approximately related to them.

4. Final measured drawing

It is important to produce accurate drawings, both in terms of content and dimensional accuracy.

Provide a north point, and information regarding scales used.

5. Equipment

(a) Measuring aids

If you are taking measurements on your own use one of the following.

(i) A measuring rod. A folding rule, which is 2 metres long when extended, will be suitable.

(ii) A retractable steel rule. Common sizes are 2 and 5 metres long.

If you have a partner to help you, a tape can be used. Thin flexible steel tapes are available, but tapes of a synthetic material are available. They are easier to use and often preferred by beginners.

(b) Sketching aids

Any of the following may be used.

(i) Stiff backed sketch pad. A4 paper is widely used, with A3 paper as a possible alternative.

(ii) A clipboard with loose sheets of A4 paper.

(iii) A 'do it yourself' clipboard, made from a stiff cardboard, hardboard, or thin plywood board, with a bulldog type clip to hold the paper in position.

Standard Self Assessment Questions

In order to satisfy yourself that your work is of an acceptable standard, after completing each drawing you should study it and check that you can answer YES to the following questions.

1 Have you drawn a border around the paper a distance of 10 mm from the edge of the paper?
2 Have you provided a standard title panel, complete with a title, drawing number, scale, date and your name?
3 Are all the lines smooth, even and of the same thickness throughout their length?
4 At positions where vertical lines join horizontal lines do they either exactly meet, or slightly cross each other?
5 Do all the dimensions scale exactly the same as their written values?
6 Are all the dimensions just above the line and in the same direction as the line?
7 Are all the arrowheads of the same size and type?
8 Have you used guide lines for all the lettering and figuring?
9 Are all the guide lines very faint—i.e. just visible?
10 Is all the lettering neat and easy to read?
11 Are all the vertical strokes of the letters exactly vertical?
12 Are all similar letters exactly the same—e.g. are all 'A's identical?
13 Are all letters of an appropriate size?
14 Are all numerals neat and easy to read with no possibility that they may be misunderstood?
15 Is your drawing of clean appearance as opposed to looking dirty and smudgy?

Bibliography

The following publications are basic sources of reference of the subject of construction drawing.

British Standards Institution

BS 1192: Part 1: 1984. *Construction drawing practice. Recommendations for general principles*.

BS 1192: Part 2: 1987. *Construction drawing practice. Recommendations for architectural and engineering drawings*.

BS 1192: Part 3: 1987. *Construction drawing practice. Recommendations for symbols and other graphic conventions*.

BS 1192: Part 4: 1984. *Construction drawing practice. Recommendations for landscape drawings*.

BS 1192: Part 5: 1990. *Construction drawing practice. Guide for structuring of computer graphic information*.

BS 1192PG: 1988. *Construction drawing practice. Graphic symbols for construction drawings*.

BS 308: Parts 1, 2 and 3: 1972. *Engineering drawing practice*.

Property Services Agency, 1989. *PSA Drawing Practice Manual*. BRE, Garston.

RIBA, 1991. *The Architect's Handbook of Practice Management* (previously entitled: *Handbook of Architectural Practice and Management*). RIBA Publications Ltd.

Tutt, P and Adler, D. 1979. *The New Metric Handbook: Architect's Journal*. Architectural Press.

Many excellent books are available dealing with the technology and other subjects allied to construction drawing, which students will find helpful in their studies and work. The following list is a representative selection of suitable books.

Reekie, F and McCarthy, A, 1993. *Reekie's Architectural Drawing* (previously entitled: *Draughtsmanship—Architectural and Building Graphics*). Edward Arnold.

Chudley, R, 1991, 1992. *Construction Technology, volumes 1 and 2*. Longman.

Newton, PH, 1991. *Structural Detailing, second edition*. Macmillan.

Burberry, B, 1992. *Mitchell's Environment and Services, seventh edition*. Longman.

McFarlane, R, 1993. *Beginning AutoCAD*. Edward Arnold.

Clancy, J, 1991. *Site Surveying and Levelling, second edition*. Edward Arnold.

Thompson, A, 1990. *Architectural Design Procedures*. Edward Arnold.

The following publications are basic sources of reference for drawing equipment.

British Standards Institution

BS 4000: 1983. *Sizes of paper and board*.

BS 1342: 1962. *Specification for detail drawing paper.*

BS 1347: 1986. *Specification for architects', engineers' and surveyors' metric scale rules.*

BS 2459: 1974. *Specification for set-squares (fixed and adjustable) for drawing office use.*

BS 5536: 1977. *Specification for preparation of technical drawings and diagrams for microfilming.*

INDEX

Abbreviations 35, 38
Active lines 31
Annotation 25, 91, 99, 143
Architectural drawings 63, 64, 134
Assembly drawings 90–98, 143
Axonometric projection 53–62, 143

Backing sheets 4, 11
Beam compasses 7
Block plan 143
Break lines 31
British Standards Specification 1, 143, 149, 150
Building engineering services drawings 65, 118–125, 135
Building line 143

Cartography 143
Cartridge paper 2
Centre line 31
Checking drawings 138
Check lists 138–142
Circle aids 7
Circles 17
CI/SfB 143
Civil engineering 143
Civil engineering drawings 64–65
Code of practice 143
Compasses 7
Component drawings 99–103, 143
Computer aided draughting 135, 136
Construction drawings 1, 63–66, 144
Construction lines 12–15, 31, 144
Contract drawings 144
Contractor's drawings 65
Co-ordination drawings 144

Datum 67, 144
Detail drawings 144
Detail paper 2
Development plan 144
Dimensioning 25–30
Dimension lines 25, 26
Dimensions 25, 26, 31, 67, 72, 78, 83, 144
Dividers 7
Door numbers 73
Doors 34, 35, 73
Door symbols 34, 35
Drain lines 31
Drawing boards 3, 4
Drawing materials and equipment 1–9
Drawing numbers 137, 138
Drawing paper 2
Drawing paper sizes 2, 3
Drawing registers 137
Drawings 1, 137–142, 144

Dye-line printing 2

Elevations 42–48, 83–89, 144
Erasers 6
Erasing machines 7
Erasing shields 6

Final lines 12, 15, 144
Floor plans 72–77
Freehand drawing 66, 126–133
Freehand lettering 23
French curves 7

General arrangement drawings 144
General layouts 118
Geometrical drawing 135
Graphical conventions 31–40
Grids 118, 144
Guide lines 24

Hatching 35, 99
Hidden lines 31

Ink 6
Isometric projections 53, 55, 56, 58, 144

Landscape drawings 135
Landscaping 67
Lettering 23–30
Lines 10–18
Location drawings 72–89, 145

Mark references 105, 107, 109
Measured drawings 66
Metric system 19

North point 35, 73
Numerals 23

Oblique projections 145
Orientation of plans 35
Orthographic projection 41–52, 145

Paper fixings 4, 11
Partitions 72–73, 78
Pencils 4, 5
Pens 6
Planning drawings 63
Planometric projections 145
Plans 41–47, 145
Plant area drawings 123
Production drawings 64, 145
Production information 126, 138–142, 145
Projectors 25, 26, 31
Protractors 7

Quadrants 17–18

Ramps 32, 35
Record drawings 66, 145
Reference points 91, 105–107
Reinforced concrete 106, 109
Reinforcement 109

Scale rulers 6
Scales 6, 19–22, 145
Schedules 145
Section lines 31
Sections 42–48, 78–82, 145
Services drawings 65, 118–125, 135
Set squares 6, 7, 11
Site plans 67–71, 145
Sketch drawings 63, 145
Staircases 32, 35
Straight edges 3, 4, ,11
Structural detailing 104
Structural engineering drawings 64–65, 104–117, 135
Structural steelwork 104–106
Sub-contractor's drawings 65
Supplier's drawings 65
Survey drawings 63
Symbols 35–37

Tee–squares 3, 4, 11
Title panels 13, 14, 15
Tracing film 3
Tracing paper 2

Walls 72–73, 78
Windows 34, 35, 73
Window symbols 34, 35
Working drawings 145